THE CONCISE ENCYCLOPEDIA OF
POULTRY BREEDS

AN ILLUSTRATED DIRECTORY OF OVER 100 CHICKENS, DUCKS, GEESE AND TURKEYS, WITH 275 PHOTOGRAPHS

FRED HAMS

southwater

CONTENTS

INTRODUCTION

What constitutes a breed? In poultry terms it has come to mean the type, or shape, and collection of genetic features that are linked closely enough to be described as a type. There are more than a hundred pure breeds of poultry in the world, many specific to geographical regions.

The diverse visual appearance and characteristics of poultry, such as the inherent ability to tolerate cold or heat, have partly developed in response to the environment in which each breed originated. They are also the result of selective breeding by farmers. Breeders would have prized different characteristics and would have developed these features in their breeding programs.

While few people would have trouble identifying poultry in general, it is likely that many of us will have seen only a few breeds in our local environment. It is the sheer variety of these birds, their differing temperaments and utility or aesthetic qualities, that appeals to poultry fanciers and exhibitors today.

Pure breeds

Nearly all of the important poultry breeds that we now think of as being "pure" have their origins in the 18th century. Pure breeds have distinct physical and visual characteristics. Later "manmade breeds" developed out of crosses between pure breeds. These incorporated various combinations of heavy-boned Asiatic fowl and the tiny ancient fowl of Northern Europe or the Mediterranean in the breeding pen. As a result,

► *In some long-tailed Japanese varieties, feather length can extend to several metres.*

modern breeds are capable of much variation. These variations will continue to evolve with the environment in which poultry live as well as from human intervention.

Bantam versions have often been developed in parallel with the exhibition strains. These now often out-number their full-size counterparts.

Few pure breeds rival the egg production of their modern hybrid counterparts, and some are kept to add beauty and completeness to gardens rather than for their output of eggs. It is wonderful that, albeit in some breeds in a miniature or bantam form, so many of the traditional breeds are still here to be studied, understood and reared.

It is these pure breeds that attract many people who see keeping poultry as a fulfilling hobby, an added bonus being a supply of fresh eggs. Today, most hobby or specialist breeders are attracted to a particular breed for its standardized show points, such as intricate feather patterns, modified silky or frizzled

▲ *Feathered feet are generally an Asian characteristic, which can vary from the outer shank feathering to fully feathered feet, where feathering extends to the middle toe.*

feathering, great height and reach or very short legs.

The term "light breed" refers to bone structure and body shape. It is derived from an earlier "sitter" and "non-sitter" classification (meaning those likely or unlikely to go broody and rear their own chicks).

Strains

Different strains of breeds also exist. A strain of a pure breed is one that has been developed by a breeder's family for generations, and has been reared and selected from a closed flock. A closed flock does not allow for any other poultry, including any of the same breed, to be introduced into the flock. In this way the bloodline remains pure through the generations, and the ancestry of the birds can be clearly traced. A strain of poultry may also refer to specific characteristics that a breeder has developed within his own flock. To the untrained eye

▶ *Modified feathering in the Naked Neck breed can extend to a complete absence of feathers on the neck.*

all strains of the same breed may look identical. After all, they all must adhere to a written standard for that breed that has been defined and approved by an officially recognized body. Strains may, however, have subtle variations such as egg-producing capacities, or some strains may consume more grass than others.

Varieties

The term "variety" is usually reserved for colour variations within a breed. There are many colours of poultry, a number of which are from selective breeding. Historically, breeds may have existed in just a few colourways, or combinations of colours. In order to introduce new colours to one breed, poultry keepers include genetic material from other breeds with the requisite shades. Colours such as Wheaten, Buff, Columbian and Silver, for example, have specific breed requirements. Birds may feature more than one colour, and the breed standard may require that a colour be specific to certain parts of the bird, or predominate in a given area. Varieties are also thus standardized, and criteria must be accepted by a relevant body.

In breeds such as Poland, variety is defined by beard or lack of beard. The standard also takes into account different feather structures. When buying imported breeds and new colour varieties with the intention of selling future offspring, check first with an accredited poultry club that the characteristics are to standard.

Hybrids

Often the birds that we see on poultry farms are known as "hybrids" rather than as pure breeds. These are fowl that are in effect artificially bred. They have been developed in response to market pressures to produce the maximum quantity of eggs for the smallest amount of feed. It was enthusiasts, rather than commercial breeders, who began experimenting with the creation of hybrids more than a hundred years ago. Through observation they determined which hens laid the largest number of eggs, and bred from those individuals. All hybrids have pure breeds in their ancestry. As the industry progressed, hybrids were created by selecting desirable features from different breeds and adding them to the gene pool. Only in a few instances do these hybrids bear any resemblance to the earlier standard-bred utility flocks.

▼ *Muffs and beards often inhibit wattle development.*

◀ *True bantams, the dwarves of the poultry kingdom, often have a low wing carriage and a jaunty stance.*

Such birds may lay 20 per cent more eggs than their pure-bred counterparts, and are ideal for many domestic situations.

How to use this book

This book is intended as a reference for those who are considering keeping chickens, or who already own poultry and wish to learn more about their own breeds or are looking to acquire new breeds for general purpose, or for exhibiting. The pages that follow provide a detailed look at more than 100 popular breeds of poultry, including ducks, geese and turkeys; some breeds with worldwide popularity.. The breeds are arranged according to type, primarily whether they are an old foundation breed used to create other newer breeds, or whether they are an artificially created crossbreed, developed from interbreeding other pure breeds. Within each category, the breeds are arranged according to their geographic point of origin in much the same way as they would be organised if exhibited at a show. The true bantams of the poultry world have their own chapter, as do ducks, geese and turkeys. An essential characteristics panel accompanies each breed, listing their key features, egg yield, temperament and housing requirements to help you decide if it is the right breed for you.

COMBS

It is the comb that distinguishes the genus *Gallus* from other bird breeds. Since the domestication of the ancestors of modern domestic fowl (all of which had combs) there have been a number of mutations affecting the visual appearance of combs.

There are a number of distinct comb types – buttercup, horned, mulberry or walnut, pea, rose, single and strawberry.

Buttercup comb

The buttercup comb is specific to the Buttercup breed of poultry. It is a fleshy comb shaped like a goblet, that sits centrally on the head and is smooth in texture.

Horned combs

V-shaped or horned combs are specific to some European breeds. The two prongs of the V are joined at the base of the comb which starts at the top of the beak. The Houdan, La Fleche, Sultan and Polish breeds all have horned combs.

Mulberry or walnut combs

The mulberry or walnut comb is small, broad and relatively flat, and sits low on the front of the head. It is

▼ *The buttercup comb is unique to the Sicilian Buttercup breed. It should look like an upturned buttercup flower.*

▲ *The mulberry or walnut comb of the Silkie is the shape and colour of the fruit.*

relatively smooth on all sides. Silkies and Yokohama breeds exhibit walnut combs.

Pea combs

The short pea comb is standardized in just five true poultry breeds, including the Sumatra and the Indian or Cornish

▼ *The horned comb of La Fleche accompanies cavernous nostrils, and occasionally a few raised feathers.*

breeds. It was first described in 1850 as being similar to a pea blossom and is a medium-length comb that starts at the top of the beak and finishes at the front of the head. Each comb has three lengthwise serrations.

Rose combs

The rose comb is clearly identifiable by its leader, a spike at the end of the comb that may, depending upon

DOMINANCE IN COMB TYPES

Much of the work into the genetics of comb types was carried out by William Bateson in 1902. He crossed a Wyandotte, which has a rose comb, with a Leghorn, with a single comb, and a Brahma, which has a pea comb. All the offspring (the F1 generation) had walnut combs, an entirely new comb type. When the F1 generation were mated, the resulting F2 generation showed a 3:1 segregation, which was said to have shown that pea and rose combs were dominant over the single combs, thus proving that Mendel's laws of genetics extended to the animal kingdom as well as plants. However, Bateson had chosen Wyandottes, which were the result of earlier unions between breeds with different combs, and this fact could have skewed the results. Comb genetics can be confusing, but fascinate exhibitors and hobbyists, some of whom make the occasional experimental cross. The fact that the purest bred Wyandottes continue to produce a percentage of single-combed offspring could again be the result of their own mixed ancestry.

▲ *Pea combs were thought to resemble the shape of an open pod of peas, or that of the pea flower.*

▲ *A rose comb may terminate in a leader (spike) that extends straight back from a wide front, or may follow the neck line.*

▲ *Single combs with even serrations are expected to stand upright in both sexes, in heavy breeds such as the Sussex.*

breed, be long, short or almost horizontal. The rest of the comb contains fleshy nodules. Several well-known breeds have a rose comb including Leghorns, Dorkings, Hamburgs and Derbyshire Redcaps.

Single combs

Most of the economically significant domestic breeds have single combs, as have *G. gallus*, *G sonneratii* and *G. lafayettii*. With the exception of the Javanese jungle fowl, all the probable

ancestors of domestic fowl have single combs. The large single comb should stand boldly upright in an exhibition rooster, while the comb of its female counterpart may flop gracefully to one side, or may also stand upright. When viewed from the side, the single comb forms a semi-circular head ornament that begins at the top of the beak and travels centrally over the top of the head, finishing at the back. Most combs have serrations, with those at the centre of the comb

standing taller than those at each end. In order to stand upright, the tall rooster comb needs a wide and strong base. To flop to one side the female comb needs to be relatively slim.

Strawberry combs

As its name suggests, the strawberry comb has the appearance of half a strawberry sitting on top of the head at the top of the nose. The Russian Orloff, for instance, has a strawberry comb.

▼ *In some breeds, the single comb should flop gracefully to one side on the hen bird, while standing upright in the rooster.*

▼ *The strawberry comb of the Malay may result from the breed's development in Cornwall, not its Malaysian ancestry.*

▼ *The Russian Orloff breed has a strawberry comb, which starts directly above the beak.*

OTHER ORNAMENTS AND APPENDAGES

Feathered feet, extra toes, crests, beards and ear muffs are fascinating, and are standardized features of many poultry breeds. Only the Sultan breed possesses all of these additional features.

Distinctive appendages such as crests and extra toes are difficult to perfect in poultry and often appeal to experienced breeders, who thrive on the challenge.

Crests

Crests that consist of feathers on the head, sometimes enhanced by an enlarged skull, seem to fascinate many of today's breeders and exhibitors. The distinctive enlarged skulls are formed by a type of cerebral haemorrhage, and are generally confined to breeds with full crests such as Poland, Sultan and some Houdans. The cranial distortion is sufficiently different in rooster and female birds to allow the sex of chicks to be determined at hatching. Single combs are rarely associated with full crests; instead comb types vary from broad and almost circular in Silkies, through leaf-shaped in Houdans, to practically non-existent in Polands.

▼ *The Houdan crest should be accompanied by a horn-shaped comb, which is small in this female.*

▲ *The profuse muff and beard of the Russian Orloff may be the result of a genetic quirk.*

Beards and muffs

These are additional feathers grouped on the head area. Found in Houdans, Faverolles, and some Polish and Belgian bantams, beards and muffs are caused by an incomplete dominant gene. Fanciers and exhibitors have selectively bred fowl to emphasize this feathery combination.

▲ *The Barbu de Watermael has a muff and triple-lobed beard.*

Vulture hocks

These appendages consist of stiff, downward-pointing feathers that almost resemble flight feathers on the heel of the tibia, at the point where most fowl have soft or downy feathers. This feature has been found to be recessive, and helps to emphasize the amount of effort breeders have undertaken to preserve and enhance appendages that may well have no utility value.

Feathered feet

While feathered feet are an Asiatic trait, there is little evidence that many of the original breed imported in the 1850s had heavily feathered feet, apart from an early reference to Burmese bantams having wing-like feathers on their feet. It fact, there is

◄ *The Appenzellar Spitzhauben crest should sweep forward like that of a Roman centurion's helmet.*

indication that the Cochins of 1855 had little more than feathered shanks. Through selective breeding, by 1895 Cochins had evolved into fowl with the feathers on both middle and outer toes that we see today in the breed. This variation of shank and foot feathering in the original Asiatic imports explains the limited amount of shank feathering on breeds such as Modern Langshans and the occasional appearance of foot feathers – a fault, in breeds like Sussex, that is inherited in part from these earlier imports.

Shortening of the outside toe (brachydactyly) is often found in those birds with the most completely feathered feet. The Sussex breeders who incorporated a percentage of feather-legged ancestry into their birds managed, by 1900, to have standardized a completely clean-legged fowl. However, breeders of the French Faverolles retained some proportion of the leg feathering, along with crests and beards inherited from Houdans, and five toes – a feature of their Dorking ancestry. While all are fancy traits, the

▼ *The Booted Bantam or Sapelpoot breed from the Netherlands has fully feathered feet.*

▲ *The pea comb and daw (pale) eye of the Asil or Aseel are still evident in its Indian Game descendants.*

Faverolles' utility strains retained many of their table qualities, which are still being selected today by hobbyists and some French farmers.

Five toes

The fifth toe (polydactyl) is an important feature of the Dorking and Silkie breeds. The extra toe must have dated from the genesis of domestic poultry keeping, though to some extent these features have been selectively encouraged and are part of the breed standard criteria. In Dorkings, judges and breeders will

▼ *Vulture hocks are stiff flight-like feathers that are attached to the hock, just below the thigh.*

▲ *The Sultan breed has all five appendages: crest, muff, beard, vulture hocks and five toes.*

expect five very well-placed toes. In breeds such as the Faverolles, the feature will have to compete with beards, and in Sultans, with a whole range of standardized appendages. Multiple spurs may be unique to Sumatra game, but as both multiple and single spurs are standardized, they are no more important than the breed's other characteristics.

▼ *Five toes are a breed standard in Dorkings. They are also found in Faverolles, Lincolnshire Buffs and Silkies.*

A DIRECTORY OF FOUNDATION BREEDS

Foundation breeds are pure breeds of poultry. They are generally old breeds with distinct characteristics and visibly identifiable features. Pure breeds are recognized officially by a poultry club. Each country has its own poultry club, which holds the written standard for each breed type. Each country's poultry club is supported by members who breed poultry to the accepted standard. Between them, these members are responsible for maintaining and safeguarding the bloodlines and breeding stock of each pure breed and ensuring that they conform to the breed standard. Members are usually passionate about their chosen breed, and each strives to produce the best poultry by selecting specific features and attributes found in individual birds, and using those birds to produce the next generation. Each pure breed must produce offspring that is a replica of its parent. Foundation breeds form the genetic basis of many of today's modern hybrid poultry.

▲ *The Sussex hen is an important foundation breed.*

◀ *Swiss Appenzellar Spitzhauben and Sumatra game are both early breeds that developed in isolation. Each will thrive in very free-range situations.*

WHAT IS A FOUNDATION BREED?

A foundation breed is considered by many poultry breeders to be any breed of poultry that has been used in the creation of another breed. Often these are old breeds of poultry that have been developed and selected over time for their significant desirable features.

For the purposes of this book the foundation breeds are characterized by geographical region – British, European, Mediterranean and Asian. These are categories that exist in show classes at exhibition, with the latter category being subdivided into soft-feathered Asian birds, and game and hard-feathered Asian birds. All of the ancient fowl groups from each of these regions provided humans with the genetic material required to create the breeds of today.

Poultry breeds differ significantly in character, appearance, egg-laying capacity and quality of meat. Even within a local area, the same general type of poultry will fare differently in the hands of different poultry keepers and farmers. Breeds of poultry were developed in response to market demand by exploiting the natural qualities of the poultry and selecting

the best birds of each generation that displayed desirable qualities from which to breed. Many of these went on to become foundation breeds.

The heavy-boned Asian fowl added significantly to the gene pool of the later economically important standardized breeds. Crosses between

◄ The Silver-laced Wyandotte was the result of some rather unlikely foundation breeds being used in formulae when American breeders first sought to make their own ornamental fowl. Later, white sports from these were used to create the White version that for a while included some of the world's best egg-laying strains.

very different Asiatic fowl produced the Asian soft-feathered Brahma breed. This in turn provided genetic material for the British Dorking-type fowl, which was then selected to create the British Sussex breed as well as a whole range of innovative American breeds and hybrid strains.

British foundation breeds

In Britain, there seems to have been significant investment in developing breed type, at first by breeders keen to take advantage of the high prices people were willing to pay for poultry with white breast meat. With the introduction of huge Asiatic birds, genetic material from Asian birds was used to develop larger table fowl with more meat on the carcass than previously recorded. The Dorking breed was one such, bred for market using genetic material from Asian birds, which added bulk to its size. Fanciers in Britain also spent considerable time and effort developing strong, light and agile birds for cockfighting.

◄ The Speckled Sussex, like the Jubilee Orpington, first developed as a table fowl out of crosses between local strains and Dark Brahmas. Their descendants developed into one of our most spectacular exhibition breeds.

▶ *Early crosses between completely unrelated heavy-boned Asiatic fowl gave the world one of its best known breeds, the Brahma. This in turn would be part of many of the world's later economically important breeds.*

European and Mediterranean foundation breeds

In past times, poultry fanciers selected and developed breeds of poultry from Europe and the Mediterranean for their ornamental value. Crests, muffs, beards, feathered feet, five toes and vulture hocks were seen as desirable traits that added distinction and character to different breeds. These breeds offered meat and eggs but were also valued for their appearance by fanciers, who were motivated by competitive exhibitions to produce the best of each type.

Early forms of European fowl were probably descended from jungle fowl· and fowl from Western Asia. Nearly all ancient European breeds are lightweight and have fabulous egg-laying qualities. Mediterranean foundation breeds differ from other European breeds in their upright stance and wing carriage, which they

share with many domesticated Asian breeds. Some types were also selected and developed for their head points. Their origins are uncertain, but heavier Leghorn-type fowl almost certainly display the influence of Asiatic stock.

Asian foundation breeds

Asian soft-feathered breeds are massive birds that have been select-ively bred for their size. However they probably no longer closely resemble the first domesticated Asian birds imported to Europe in the 19th century, an action which greatly influenced the development of many modern breeds. The hard-feathered breeds of India and South-east Asia have also influenced the development of many modern poultry breeds outside the region.

▲ *Table-type New Hampshire has similar coloration to the Rhode Island Red which probably results from both having some Red or Cinnamon Malay in their founding ancestry.*

HOW TO USE THE DIRECTORY

Each poultry breed listed in the following chapters of the book is accompanied by an account of breed history and interesting qualities. Each has an Essential Characteristics panel, which contains a number of features. The desirable weight of standard and bantam versions (where applicable) are given. The panel also lists the environment in which the breed thrives, the breed's temperament, and the number of eggs a good specimen of a utility breed might produce. NB There is considerable variation in egg numbers produced by different strains of a breed. In many cases, exhibition strains lay very few eggs, since other characteristics are deemed more important to the breeder. Weights given for bantams are often exceeded by show winners. The general rule is that if no bantam weight has been agreed, most individuals will be expected to weigh between 20 and 25 per cent of the large version.

▼ *Utility-type Rhode Island Red rooster (below) running with Light Sussex hens (right) would have made up much of the early British laying flock.*

BRITISH FOUNDATION BREEDS

In Britain, the selection of birds for cockfighting historically influenced the types of fowl that were bred, as well as those physical features that were considered desirable traits in fighting birds. Generally, such birds have a lightweight appearance and flighty nature.

Later, with the introduction of heavy-boned Asiatic breeds, heavy poultry breeds were developed to meet a growing demand for table fowl. It is likely that many native British foundation breeds were influenced in their breeding by Asiatic fowl. Such birds naturally grow bigger than lightweight fowl. Prior to the introduction of hybrid lines to the poultry industry, many British pure breeds were also selected for their white flesh and plump breast meat, which was highly saleable.

The foundation breeds that appear in the following pages have all been developed by British breeders for specific character traits. Many are widely known all over the world. Others have retained their regional popularity and remain contained within a small geographical area. Such birds are in the hands of a few breeders. All of these fowl are considered heritage breeds, with some being rare and highly prized.

Hamburg/Hamburgh

Hamburgs are known to belong to an ancient group of fowl that are thought to have been dispersed along the routes of Viking conquests or Nordic traders. Contrary to its name, the Hamburg breed is thought to have originated in the Netherlands rather than Germany, though it may have passed through the port of Hamburg on its way to Great Britain. The breed, as we know it today, was refined by British poultry fanciers in the mid-19th century, at a time when the first poultry craze was sweeping the country. The standard for the breed was set at this time.

It is because British breeders refined the breed that the Hamburg is listed as a British breed.

In appearance, Hamburgs have a light bone structure and almost pheasant-like carriage. They are a soft-feather breed and a prolific layer of white eggs. Neither large or bantam versions lay a big egg, although both are capable of converting a given weight of poultry feed into as great a weight of eggs as any traditional breed. All, or nearly all, have a rose comb, white lobes and slate-coloured

ESSENTIAL CHARACTERISTICS

Size: Large male 2.5kg/5lbs. Bantam male 680–794g/24–28oz. Large female 1.8kg/4lbs. Bantam female 624–737g/22–26oz.

Varieties: Black, Gold-pencilled, Gold-spangled, Silver-pencilled, Silver-spangled, White.

Temperament: Flighty, active.

Environment: Happiest given free range.

Egg yield: One of the best layers of small to medium, white to tinted eggs.

legs. Typically, they have either a gold or silver ground colour with some form of black pencilling or spangling. Prior to the breed standard being established, local strains in northern England were selected for breeding refinement on the basis of the precision of their feather

▶ *The Hamburg breed has a shared ancestry with much of the poultry indigenous to Northern Europe and the Eastern Mediterranean. Regional populations of the Hamburg breed were developed with different traits. This is a bantam Silver-pencilled hen.*

▶ *A standard Black female Hamburg.*

patterns. Strains known locally as Bolton Greys and Bays were selected for the quality of their pencilling, which took the form of a series of fine straight lines across the feather. Similarly, Lancashire Moonies, another strain, were selected for their precise spangling to each feather end. Other regional British strains included varying forms of Redcaps and Pheasant Fowl, which were known locally as Corals, Creels and Chitterpats.

Regional Mediterranean populations, in contrast to British birds, often have either broken barring or spangling on the feathers. Because of the similarity of many of these local populations, organizers of the first shows chose to categorize all of them as Hamburgs. This decision later caused countless complications for those charged with providing a written standard of the breed. Gold and silver varieties, for example, may have different markings to those with a different ground colour. However, such a wide definition of the breed has proved advantageous for its conservation, as it ensures that other bloodlines have not been used in its make-up and that the breed retains genetic vigour. As it is difficult,

▼ A large Gold-pencilled male. The Hamburg is a hardy breed that can cope with cold weather.

thanks to their exacting colour standards, to make use of any out-cross, the genotype preserved within many Hamburgs remains virtually identical to that of fowl that existed a century ago.

Because breeders pursue precise feather marking, many Hamburg bantams are permitted to creep up beyond the standard bantam weight. Most of the large versions fall just short of their standard weight. Black bantams have never been standardized in Britain, as the Rosecomb bantam breed is thought to be too similar in appearance to warrant a separate classification. Breeding exhibition-quality examples or show winners without the help of someone knowledgeable may prove difficult, and breeding stock of the large versions may be hard to find, but the bantam versions are generally plentiful.

These birds are flighty by nature and, given their desire to range, will need to be surrounded by high fencing. They are active foragers, lively and alert, and do not do well in confinement. Their temperament makes them perfectly suited to barnyards, though they can make appealing garden companions.

▶ The Hamburg also shares its ancestry with some of the early English game strains. Being reared in a harsh climate helped to produce a breed with an in-built toughness that was invaluable to breeders. This is a bantam Silver-spangled female.

▶ A large Silver-pencilled male.

Old English Pheasant Fowl

This is a distinct regional breed from Yorkshire, Lancashire and the farms of Cumberland. It is an old breed, named in 1914, and has the Hamburg in its lineage. Unlike the fanciers and exhibitors who selected the colours and feather patterns of the Hamburg breed to conform to exacting standards, farmers and commercial poultry keepers selected their fowl on the basis of vigour, egg-laying ability and quality of meat. The result is the Old English Pheasant Fowl, once known as the Yorkshire Pheasant. Good examples are still seen in the rare breed classes at exhibitions. For much of the 20th century there was enough support for the breed to make it a commercially viable alternative to the main poultry breeds. As a farm bird from Northern England, it is known to be tolerant of cold, and is still highly regarded as a dual-purpose bird. The hen is a good egg-layer and the male is kept for its meat.

The basis of the breed's colour standard is a simple black crescent at the end of gold or silver feathers. It is a graceful bird with pheasant-like characteristics.

It has a rose comb, white earlobes, and slate-coloured legs and feet.

ESSENTIAL CHARACTERISTICS

Size: Large male 2.7–2.9kg/6–7lbs. Large female 2–2.7kg/4½–6lbs. Bantam male 794g/28oz. Bantam female 737g/26oz.
Varieties: Gold, Silver.
Temperament: Active and alert but can be quieter than the related Hamburg strains. Hardy.
Environment: Suits a free-range environment.
Egg yield: A good number of white eggs per year.

◀ An Old English Pheasant Fowl is a forager by nature.

Derbyshire Redcap

True to its name, the Derbyshire Redcap originated in the county of Derbyshire, and was recorded as early as 1848, though the breed had to wait for national recognition. The breed's most distinguishing feature is its comb, which is made up of fleshy protrusions that cover the head. The comb can be up to 7.5cm/3in long and points backward. Initially, emphasis was placed on the size of the comb, which may have detracted from the overall usefulness of some strains of the breed. However, the standard defines the quality, rather than size, as being the breed's most important feature. The fact that it has red, rather than white, earlobes sets the breed apart from its closest relatives. The bird is a light breed with soft feathers, and has a single body colour which varies from orange to brown. Its tail plumage is black and its legs and feet are slate.

This bird has an active nature, making it suited to a free-range environment. It was originally kept as a dual-purpose bird.

▶ This Derbyshire Redcap male has black plumage.

ESSENTIAL CHARACTERISTICS

Size: Large male 3.4kg/7½lbs. Large female 2.7kg/6lbs. Many large fowl do not reach this weight. Bantams weigh 20–25 per cent of large fowl.
Varieties: Black.
Temperament: Active. A keen forager.
Environment: Free range. Needs a large amount of space.
Egg yield: 150–200 large white eggs per year.

Dorking

Named for the town of Dorking in Surrey, this is an ancient breed of poultry with Italian ancestry, thought to have originated at the time of the Roman Empire. The breed is prized for its five toes, an unusual feature in poultry. Five-toed poultry were recorded by one Roman author, and so there is an assumption that the breed was taken to the Dorking region by Roman legionaries. The early Dorkings depicted in art are slim, game-like fowl with five toes.

This is an important breed with a significant place in poultry history. The breed was highly prized in the mid-19th century, when white flesh was thought to indicate delicacy of flavour. By the 1860s, the Sussex breed had replaced the Dorking breed as the poultry of choice for the London markets. At this point, the French used the Dorking to help create the Houdan and Faverolles breeds that were sold through the Paris poultry market. The Dorking breed remained popular until the mid-20th century, when farmers began to change their flocks to hybrids.

During the mid-19th century, the Dorking standard, breed type and colours were stabilized and settled in a form that can still be seen among the best examples at major shows today. Standardized in five colours, Silver-grey, Dark, Red, Cuckoo and White, the breed also has single and rose combs, though not necessarily in each of the standard colours. The white variety is often found to be smaller than birds in other colours. The rare Cuckoo variety is always expected to have a rose comb, while the Red variety and the common Silver-grey, by contrast, should have only a single comb. The Dark variety, with its enigmatic

▼ *This is a Silver-grey pullet, smaller and slimmer than the hen.*

coloration, which seems to owe much to the old Kent and Surrey regional strains of fowl, may have a single or rose comb. The breed has a square frame, short, clean legs and five toes. It is a large breed, with males weighing up to 4.1kg/9lbs, and so requires a large poultry house and plenty of space. The standard for weight varies considerably between countries, and the British standard contains the heaviest weight, though not all birds attain this. This is a hardy breed, although the five toes and the male's large comb may need protecting in winter. The female is an average egg-layer, but interestingly, this breed is the only one with red earlobes to lay a white egg. The breed also has a red comb and wattles, and white legs. It can cope well with confinement, and is shy and docile by nature, making it the perfect companion in a garden. For flavoursome meat and eggs this breed must be fed good-quality feed. Bantam versions of the old single original breeds have proven difficult to perfect and are becoming rarer.

ESSENTIAL CHARACTERISTICS

Size: Large male 4.1kg/9lbs. Large female 3.2kg/7lbs. Bantam male 1.1–1.3kg/40–48oz. Bantam female 0.9–1.1kg/32–40oz.

Varieties: Coloured, Silver-grey, White.

Temperament: Calm and docile.

Environment: May require some protection from cold.

Egg yield: 150 large cream eggs per year.

◀ *Anyone interested in studying the character and evolution of British foundation poultry breeds could well find a starting point in the Dorking breed. This is a large Silver-grey hen.*

Sussex

The breed we know today as the Sussex developed from the indigenous fowl of the south of England – birds once recorded as Kent and Surrey breeds. For much of the 19th century the old breed had an almost unchallenged reputation for its quality poultry meat, which was sold to the important London markets, quickly surpassing the Dorking breed as the poultry of choice for the table.

This early market required very large-framed birds. South-east England had long been home to a population of large-bodied, white-fleshed fowl, and thus many breeders would have used this stock for the basis of their breeding pens. Soon, adventurous breeders began introducing the newly imported Asian Cochin poultry into the breeding pens in order to produce larger fowl. The Sussex that were later standardized as a pure and distinct breed, separate from the Kent and Surrey breed, are made up of regional indigenous fowl crossed with Asiatic imports. Three varieties of Sussex were originally recorded – Speckled, Brown and Red – although eight are now standardized. During the whole of the pre-standardized period, the poultry would have been scrutinized by poultry dealers and consumers. Such a high level of scrutiny probably resulted in all varieties of Sussex being

▲ *A large Speckled Sussex hen.*

true enough to their English roots for the breed to be considered one of the poultry world's foundation breeds. The Sussex breed that we know today dates from around 1900, though it existed and has been husbanded commercially for at least 50 years prior to that date. In 1902, the exhibition standard for the breed was established. Like most of the 20th century's successful breeds, the Sussex has proven capable of development to meet the demands of changing times and circumstances.

As a growing middle class wanted a more reasonably priced table fowl, the Light variety was developed in order to lay the extra egg numbers needed to produce chicks for fattening for market. The Sussex

---

ESSENTIAL CHARACTERISTICS

Size: Large male 4.1kg/9lbs. Large female 3.2kg/7lbs. Bantam male 1.1kg/40oz. Bantam female 794g/28oz.
Varieties: Brown, Buff, Light, Red, Silver, Speckled, White.
Temperament: Alert and docile.
Environment: Free-range or confinement.
Egg yield: 200–260 eggs per year.

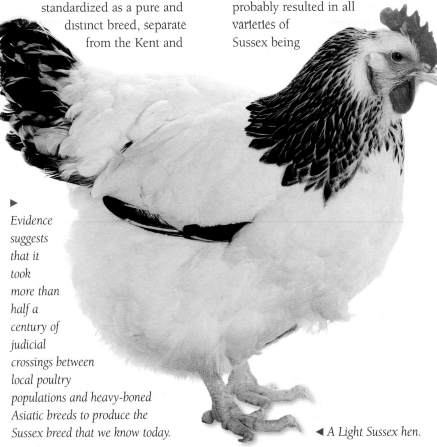

▶ *Evidence suggests that it took more than half a century of judicial crossings between local poultry populations and heavy-boned Asiatic breeds to produce the Sussex breed that we know today.*

◀ *A Light Sussex hen.*

FATTENING FOR MARKET

Historically, in order to produce birds for market, male birds were surgically caponized. After a period of captivity, if the birds had not reached the size and weight required for market, they were force fed with a mixture of locally grown ground oats, imported condensed milk and even tallow – a practice regarded as cruel today.

▲ *A prize-winning large Speckled Sussex rooster.*

▼ *A bantam Silver Sussex rooster.*

breed were selected for their white legs and flesh. When the variety was later selected as an egg-laying fowl, it consistently produced enough eggs to be considered as a utility fowl; it became one of the world's best heavy-breed egg-layers.

Today most strains of the large versions, including the later Silver and Buff varieties, are now kept as exhibition fowl. Meanwhile, the bantam Light variety has become one of the most popular and successful exhibition breeds of all time. The breed owes much of its popularity to the fact that the Silver hens, when crossed with Gold males, produced sex-linked chicks, meaning that the sex of the chick can be determined by colour on hatching. This provides an important economic motive for keeping the breed.

In appearance the Sussex is a large and graceful bird, with a broad, flat back and a stocky appearance. Its tail should be held aloft at a 45-degree angle. It has red earlobes and dark orange or red eyes, but white skin and legs. It has a docile nature and can adapt well to different surroundings. Today it retains its dual-purpose capacity. While some strains of old table-type Lights can lay more than 200 eggs, and a few Whites may lay more than 250 eggs per year, some exhibition strains of Lights are poor layers.

◄ *The Sussex breed absorbed related local strains into its make-up, as well as just enough Asiatic blood to produce a fowl that served a demanding meat trade, while at the same time conforming to established standards for the breed. This is a bantam Light Sussex rooster.*

◄ *A bantam Brown Sussex rooster.*

Scots Grey

The tall Scots Grey breed is at least 200 years old and is possibly older, since identifiable related populations are known to have been kept as far back as the 16th century.

Appropriately named, this breed hails from the crofts and farms of Scotland. It also has a dark grey base colour, a coloration that could suggest some early connection with the pencilled grey fowl of Northern England. In appearance, the Scots Grey is a tall, upright bird with long legs, a trait that was selected at a time when exhibition classes for game birds were dominated by tall examples. The breed has white skin, a single comb and red earlobes. The very full tail is essential for the standard of perfection and a significant part of the breed's character. This feature suggests that grey game may also be part of its genetic composition. This is a naturally active breed that prefers a free-range, foraging life. The inter-war years saw the breed find favour among farmers

▼ A large Scots Grey hen is found in only one colourway. It is suitable for those looking for an active fowl capable of adapting to a challenging environment.

looking for an all-round fowl that would lay a reasonable number of white eggs without the need for large quantities of purchased food. Today it is still classed as a dual-purpose bird, although the Scots Grey is also an endangered species and is largely kept for exhibition purposes. Large and bantam versions are available. The breed has retained a loyal following in Scotland.

ESSENTIAL CHARACTERISTICS
Size: Large male 4.1–5kg/9–11lbs.
Large female 3.4–4.1kg/7–9lbs.
Bantam male 623–680g/22–24oz.
Bantam female 510–567g/18–20oz.
Varieties: Barred, Cuckoo.
Temperament: Active.
Environment: Free range.
Egg yield: May lay more than 200 large white eggs per year.

Scots Dumpy

The Scots Dumpy, as its name suggests, sits low to the ground on very short legs. It has a boat-like appearance, with a broad and flat body. Its appearance is that of a heavy bird. This breed has a single comb, with red eyes and earlobes. It was kept originally as a dual-purpose bird, laying 180 eggs per year. Because the bird did not exercise much, the meat was highly regarded as it was probably wrongly thought to have less tough leg meat. Its shortened legs ensure that the bird waddles from side to side, no doubt an attractive and appealing characteristic to farmers keen to have birds that would not roam far. Its docile nature also ensured that flocks could be caught

▼ A Scots Dumpy bantam pullet, in one of the new, but as yet non-standardized, colours.

▼ For years Cuckoo was the most popular colour in large and bantam Scots Dumpy birds.

ESSENTIAL CHARACTERISTICS
Size: Bantam male 794g/28oz.
Bantam female 680g/24oz.
Varieties: Black, Blue, Cuckoo, White.
Temperament: Docile.
Environment: Free range, but readily adapts to close confinement. Unsuitable for muddy conditions.
Egg yield: Up to 180 white to dark brown eggs per year. Some strains of bantams lay surprisingly large eggs.

easily. For exhibition purposes, the Dumpy is listed as a light breed. Light breeds are generally more productive birds than heavy breeds, and are therefore most likely to be kept for eggs than meat.

Examples of dumpy birds, or fowl with extremely short leg bones, have been found among archaeological excavations. This breed has also been known as "creepies", "bakies" and "crawlers" in response to its short leg length. An analysis of excavated Dumpy bones found in York, dating from the 10th and 12th centuries, was carried out by Dr Enid Alison. In her report she found that 31 per cent of those with reduced-scale leg bones were male, and 24 per cent were female bones. Along with these bones were others of varying leg length, assumed to be from the same species. Several other short-legged breeds exist around the world, including the white-lobed German Kruper and the even shorter black-legged Danish Waddler, so it can be assumed that as a breed and gene type, shortened leg bones are a long-established part of the poultry genotype. Today two versions of the Scots Dumpy exist, and so it can be assumed that

▼ *A large Black Scots Dumpy hen.*

this breed is extremely old. The long-legged Dumpy has legs that measure 6–7.5cm/ 2½–3in long, and the short-legged version has legs that measure just 4cm/1½in. This short-legged version has a lethal creeper gene, similar to that in Dexter cattle. If two short-legged birds are mated, 50 per cent of the eggs will hatch as short-legged chicks, 25 per cent will hatch with longer legs, and the remaining 25 per cent will fail to develop as a viable embryo, or will die upon hatching. In order to reduce the prospect of such a large quantity of eggs that fail, a long-legged Dumpy can be bred with a short-legged Dumpy. This will result in half of the

▼ *A Buff Columbian Scots Dumpy bantam female.*

brood having long legs and the other half having short legs. The long-legged Dumpys are not affected by the gene.

Large and bantam versions of this small breed can be found. Cuckoo, Black and White varieties are common in the large version, but plenty of bantam sports are known. That we now have bantams in a wide range of standard colour patterns may help to cement the breed's future.

Unlike the Scots Grey, this breed did not have a special interest club until 1993, a factor that could explain why many exhibition examples were bred to similar feather patterns as the Scots Grey. In the 1920s, the Scots Grey Breed Club secretary is known to have complained that the breed had lost many of the colour varieties seen in the late 19th century. Dumpy hens make excellent mothers and often go broody.

▼ *A Black Scots Dumpy bantam female.*

EUROPEAN FOUNDATION BREEDS

The earliest forms of European domesticated fowl have a shared ancestry with those of Western Asia, as well as with one or more species of jungle fowl. Red jungle fowl look similar to the native fowl of Northern Europe, with their characteristic light body weight and more agile nature. This group was distributed further afield along the Danube Valley and Atlantic trade routes that were exploited by Nordic and Viking explorers.

Nearly all of the ancient fowl and regional breeds of Europe are now classified as light in weight, and are significant because of their egg-laying capacity. The older members of this group probably have a shared ancestry with both the very lightweight Nordic Hamburg-type fowl and early crested types. Crested types appear to have had ornamental value for much of their known history. For example, the Swiss Appenzellar has a small, forward-sloping crest and horned comb, seen in early Roman poultry sculptures, combined with the typical Nordic light body shape and rudimentary spangling. At the other end of the size scale are some of the more statuesque Mediterranean Leghorn-type fowl, thought to be the result of an early introduction of Asiatic bloodlines.

Appenzellar Spitzhauben

The horn comb and forward-pointing crest are unique features of the Appenzellar Spitzhauben. Its body type, tightly packed body feathers and colouring of black markings on silver or gold ground leave no doubt of its relationship to breeds native to North-west Europe and the Eastern Mediterranean. Poultry populations that once existed in different localities as a patchwork of strains, such as the Yorkshire Hornet breed, which became extinct in the 1930s, were almost identical to the Spitzhauben.

"Spitzhauben" translates as pointed bonnet, a reference to its head feathers. A second variety from Appenzell in Switzerland, known as the Bearded Barthuhner, suggests there is local diversity within the breed's homeland. Those who reared the local fowl of the Appenzell district benefited from the region's remoteness in maintaining the purity of the breed. By temperament the breed is intolerant of confinement, and has a

ESSENTIAL CHARACTERISTICS

Size: Small with very lightweight frame. No bantams.
Varieties: Black, Chamois, Gold-spangled, Silver-spangled.
Temperament: Flighty, active.
Environment: Happiest given the most free range conditions.
Egg yield: Can exceed 220 medium or small white or tinted eggs.

▲ *The head of this male shows a well-developed horn comb. The forward-pointing crest is unique to the breed.*

lively nature well suited to living in trees and the mountainous terrain of its homeland.

This breed has the capacity to convert reasonably small amounts of food into numerous medium-sized eggs. It is now a rare breed; in Britain it is officially classified as rare, but international breeding groups are kept by conservationists in Switzerland and other European countries. Their alert appearance and attitude helps make them look far better in the open than in a show pen, and so this is a breed that appeals to the conservationist rather than the committed exhibitor.

▶ *A Silver-spangled Appenzellar rooster.*

Sultan

The very ornamental Sultan breed with its feathered legs and feet, which are more usually associated with Asiatic breeds, has sufficient characteristics to place it in the European group of crested poultry breeds. Look closely at the Sultan and almost every deviant from the genetic norm will be found – a combination found in no other breed. Feathered feet apart, the stiff hock feathers usually described as vulture hocks, the full crest, horned comb and cavernous nostrils are all features associated with regional breeds of Western Europe.

The Sultan came to Britain from Turkey in 1854 at the height of the first poultry-keeping boom at a time when interest in everything Asiatic was heightened following the importation of Chinese Cochins for Queen Victoria. From Britain the birds were taken to the USA. A single batch of birds was imported from Constantinople, and with such a tiny genetic base the breed was destined to remain rare, appearing fleetingly at

◄ Sultans are kept primarily as exhibition birds. They have a quiet disposition and can tolerate confinement. This is a large White rooster.

▼ The Sultan has every additional feature including a flamboyant crest, long tail, horn, muffs, beard, feathered feet and a fifth toe. It can also fly. This is a large White hen.

shows for the next century until early in the 1970s, when it was chosen as the breed logo of the newly formed British Rare Poultry Society. Almost all of the examples seen over the next decade were poor specimens lacking essential features. Eventually, one breeder working in Holland, Dr Boks, virtually recreated the breed by crossing and re-crossing breeds that each displayed one of the required features. At the same time he

included the Breda, the duplex comb of which appears, at first sight, to be at odds with the requirement for the Sultan to produce a horned comb. Such is the complexity of the relationship between comb and crest that the Breda's duplex comb was an essential part of the genetic mix. Stock imported from this single source provided most of the really good examples seen over the next two decades. The breed is still available but requires the attention of dedicated breeders, who will find that its exhibition features make it one of the most ornamental of fowls.

▼ The Sultan's horn is usually covered by white feathers.

ESSENTIAL CHARACTERISTICS
Size: Large male 3.7kg/6lbs. Large female 1.8kg/4lbs. Bantam male 680–793g/24–28oz. Bantam female 510–680g/18–24oz.
Varieties: Black, Blue, White.
Temperament: Docile, friendly.
Environment: Copes with confinement well.
Egg yield: Can exceed 180 white eggs per year.

Poland/Polish

Breeds standardized as Poland or Polish, with a domed skull enhanced by full crest, were recorded as long ago as 1600. The remains of a fowl with a domed skull were found on a Roman altar site at the village of Uley in Gloucestershire, suggesting that the Romans knew the breed type and possibly held it in some esteem. The characteristic domed skull is in fact caused by a form of cerebral haemorrhage, and sets apart the fully crested breeds from related types that have varying amounts of feathers on their heads.

▼ A frizzled large Blue Poland. Frizzled feather types are less common than standard feather versions.

> **ESSENTIAL CHARACTERISTICS**
> **Size**: Male 2.7kg/6lbs.
> Female 2kg/4½lbs.
> **Varieties**: Black-crested White, Buff-laced, Golden, Silver, White-crested Black, White. Bearded or Non-bearded.
> **Temperament**: Quiet.
> **Environment**: Tolerates confinement. Requires protection in winter.
> **Egg yield**: 80 per year.

Many breeds that may or may not have crests, but which have unusual comb forms, have cavernous nostrils. With 2000 years of development of this sort of fowl, it would be surprising if some of these traits had not been incorporated into other European regional breeds that would have developed alongside them. Neither is it surprising that a breed that had developed in more than one location had also acquired a number of names and differing descriptions. Recording the breed's homeland as Padua, Italy, Renaissance author Aldrovandi called them Paduan fowl, but crude woodcuts that have survived from that region show a female with crest muff and beard, and a male with full crest and wattles. The breed has a linked gene that suppresses wattles on bearded birds. None of the known names given to the breed have any connection with Poland or Polish communities. It is most likely that the name is derived from their "polled characteristics", a mutation that means that where the best crests are found, those birds have no combs.

Twelve colour variations are known for this breed, and all the colours are available as frizzles. The bird does not go broody and lays few eggs. These small birds make attractive and endearing pets, though they are more suited to people with experience of poultry keeping. The domed crest may obscure the bird's vision, and it may also hide pests that can affect the health of the bird.

▲ Though known as Poland or Polish this breed is not thought to originate in that country. This is a White-crested black hen.

Many of the birds that were the foundation of the Poland breed in the UK were imported from the Netherlands, where they were referred to as Polder fowl or Polderlanders. Other Dutch breeds that would have been far closer to Campines and Hamburgs were referred to in the same way. In the UK, bearded and non-bearded varieties are treated as one breed. However, much of the rest of the world treats them as different breeds or as varieties of the same breed. Some of our best strains descend from stock imported from Holland during recent years. The breed is a healthy and productive egg-layer. It is generally intolerant of wet conditions, and anyone wanting to keep this beautiful, ornamental breed at its best should provide housing with a covered run or open-fronted loose box.

The size of the popular bantam versions may be an issue to some. With crest size being an important exhibition feature, some winning bantams have very large crests that appear to be oversized for bantams.

▶ The Poland is a visually distinctive breed, with the large crest providing the defining characteristic. This is a Blue hen.

La Fleche

Once raised as a dual-purpose bird and laying up to 180 eggs per year, La Fleche is now a rare breed that is kept for ornamental purposes. This bird hails from France and is thought to have Black Spanish and Crevecoeurs in its ancestry. Despite its black plumage, legs and feet, the flesh is white. The spectacular twin-horned comb that is the defining feature of

La Fleche is always accompanied by cavernous nostrils and occasionally, though not a standard requirement, a few crest-like head feathers. Its other distinguishing features are the white earlobes and long red wattles. Just as beards usually suppress wattle

development, very large, well-developed horn combs are rarely found on those birds that have fully developed, enlarged crests and skulls.

This is an active breed that likes to forage – a heavy bird that stands tall and proud. A bantam version of the breed is known.

ESSENTIAL CHARACTERISTICS
Size: Male 3.6–4.5kg/8–10lbs.
Female 2.7–3.2kg/6–7lbs.
Varieties: Black, Blue, Cuckoo, White
Temperament: A restless bird.
Environment: Copes with confinement. Can fly so needs a high fence.
Egg yield: 180 white eggs.

◀ The distinctive feature of La Fleche, a breed that is less commonly seen at exhibition, is the horned comb.

Campine

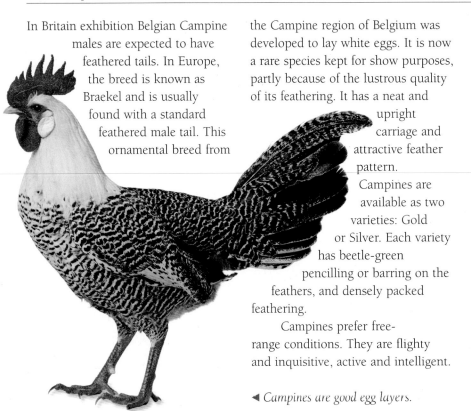

In Britain exhibition Belgian Campine males are expected to have feathered tails. In Europe, the breed is known as Braekel and is usually found with a standard feathered male tail. This ornamental breed from the Campine region of Belgium was developed to lay white eggs. It is now a rare species kept for show purposes, partly because of the lustrous quality of its feathering. It has a neat and upright carriage and attractive feather pattern.

Campines are available as two varieties: Gold or Silver. Each variety has beetle-green pencilling or barring on the feathers, and densely packed feathering.

Campines prefer free-range conditions. They are flighty and inquisitive, active and intelligent.

◄ *Campines are good egg layers.*

ESSENTIAL CHARACTERISTICS
Size: Large male 2.7kg/6lbs.
Large female 2.3kg/5lbs.
Bantam male 630g/24oz.
Bantam female 567g/20oz.
Varieties: Gold, Silver.
Temperament: Active, social.
Not appropriate pets.
Environment: Free range preferred.
Egg yield: Some early utility strains may have laid in excess of 220 medium-size eggs per year but exhibition strains may lay far less.

The breed rarely goes broody. In its original, less modified form it was used to make the original Cambar auto-sexing breed.

Lackenvelder

One of the better known breeds of the European foundation group, the German Lackenvelder enjoyed some popularity before World War II. Its name is Dutch and derives from the translation of white linen (*laken*) spread over a black field (*veld*). The breed was refined in Germany. Only one standard is recognized for this breed, although other colour variations are known. It has a black head and shoulders and black tail plumage, with the balance of the body feathers being white. The distinctive coloration is continued on to the large red comb, which always has five spikes. The legs are slate-coloured and the eyes are a dark red-brown. Like nearly all of the European foundation group, the Lackenvelder is a prolific layer of reasonable-size eggs. It is a medium-size bird that is a good forager. It prefers to free range, but can tolerate confinement. As well as eggs, the breed is kept for exhibition.

The breed is not closely related to the similar, but slightly heavier, Buff Vorverk breed.

ESSENTIAL CHARACTERISTICS
Size: Male 1.8–2.3kg/4–5lbs.
Female 1.4–2kg/3–4½lbs.
Varieties: Blue.
Temperament: Wild, flighty.
Environment: Free range preferred.
Egg yield: 200 small to medium white or tinted eggs per year.

◄ *A Lackenvelder male with some black feathers in its saddle hackles.*

Thuringian

Hailing from the Thuringian Forest region of Germany, this is an old breed known to have been around for at least two or more centuries. It is now a rare breed, even in its home country. Originally it was bred as a dual-purpose bird, with the hens laying an average of 160 eggs per year. Locally it was known as "chubby cheeks" because of the feather shaping at the sides of the face. This breed is cold-tolerant, lively and attractive. Thuringians are small birds, also available as bantams, and as they make good foragers, enjoy a free-range environment.

Gold- and Silver-spangled varieties are the most common, though plenty of others have been seen. This prettily marked breed is one of several similar local European breeds and varieties which, with closer contacts with

breeders and shows, are likely to find their way to Britain. While each breed will bring something new to the show scene, each new arrival will have to recruit its own band of breeders

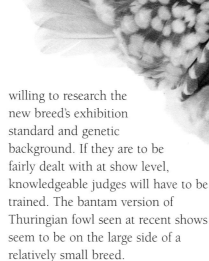

▼ *A Chamois bantam hen.*

willing to research the new breed's exhibition standard and genetic background. If they are to be fairly dealt with at show level, knowledgeable judges will have to be trained. The bantam version of Thuringian fowl seen at recent shows seem to be on the large side of a relatively small breed.

▶ *Thuringians have a small beard and single comb.*

ESSENTIAL CHARACTERISTICS

Size: Large male 2–2.5kg/4½–5½lbs. Large female 1.5–2kg/3–4½lbs. Bantam male 737g/26oz. Bantam female 623g/22oz.
Varieties: Black, Blue, Cuckoo, Chamois, Gold-spotted, Partridge, Silver-spotted, White, Yellow.
Temperament: Friendly.
Environment: Free range.
Egg yield: 160 white eggs per year.

◀ *Thuringians are small birds with a distinctive upright carriage. This is a Silver-spangled bantam rooster.*

Vorverk

This breed was created in the early 20th century in Germany by a breeder keen to establish a bird similar to the Lackenvelder but which had a dual-purpose role. Lackenvelder and Buff Orpington are two of the breeds that were used in this breed's selection, and in fact, the Vorverk appears as a black-buff version of the Lackenvelder, with black neck, collar and tail but with a buff body. The breed is rare today and is mostly kept by those interested in preserving a heritage breed. Considering its progenitors the breed should be heavier, and a newly-made bantam strain has added to confusion about the breed's size and identity. The bird is an active forager that converts feed to eggs economically. It has a rounded body, single serrated comb and orange-red eyes. The breed has a placid and alert nature.

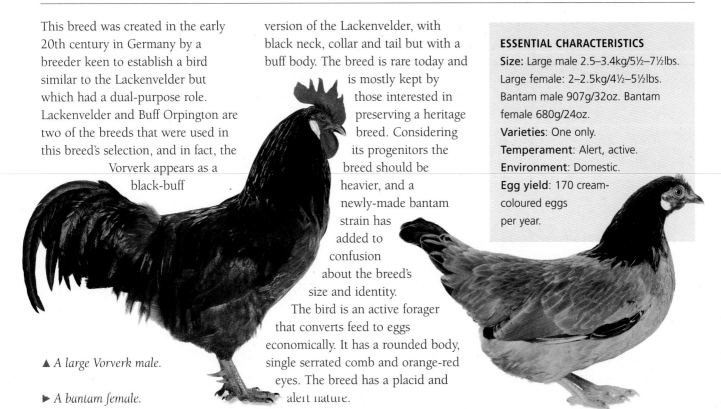

▲ *A large Vorverk male.*

▶ *A bantam female.*

Friesian

This most attractive bird is an ancient breed from the Friesland area of Holland. It is classed as a large light bird, but is almost a bantam in size. Males typically weigh 2.5kg/5½lbs and females 2kg/4½lbs. For such a small size, this bird lays large eggs. All colours have slate-blue legs and, notably, white earlobes. It is this characteristic for which it has historically been valued. The breed is known to lay early and, for its size, lays masses of eggs. It is known to be a nervous and flighty bird, which does not cope well with confinement. It is available in several colourways including Chamois, Pencilled, Gold-pencilled and Silver-pencilled. These pencilled feather patterns give some indication of a genetic link to the similarly marked, but far larger, Campine and Braekel breeds.

◀ *This Gold Friesian female demonstrates the visual appeal of the breed.*

MEDITERRANEAN FOUNDATION BREEDS

Mediterranean breeds are assumed to be those kept in Italy and Spain. However, the origin of this group remains something of a mystery. All of the Mediterranean breeds now known and kept were developed and bred to an agreed standard, and in most cases, they were named outside their country of origin. Dutch and English breeders or fanciers first chose to select breeds of the Iberian Peninsula for "head points", for example. As with so many regional populations, poultry breeders have been able to develop and capitalize on the genetic qualities that centuries of local development have provided.

While the ancient breeds of the rest of Europe have blue-grey legs and a similar body shape to red jungle fowl, Mediterranean and Iberian breeds have the upright stance and wing carriage seen in many of the domestic strains and breeds of Asia. The Leghorns and other Italian breeds also seem to have absorbed many features of other Old World groups. For instance, many Mediterranean breeds no longer have the broody traits that inhibited year-long egg production, a feature of other breeds that have a small amount of Asiatic blood in their genetic make-up.

Minorca

The Minorca breed was popular in the west of England. A single importation was taken to England from Minorca and became known in England as the Red-faced Spanish breed, a name that lasted for more than 150 years because of its similarity to the Spanish breed. With the decline of the Spanish breed, by the 1870s, the Minorca breed filled large classes at shows, competing for attention with huge Asiatic breeds and newly created Modern Game. In order to increase the size of the breed, breeders started to introduce unrelated genetic material to the Minorca gene pool. In addition, there was a perceived need to establish a distinctive wedge-shaped body and sloping back as a standard.

Over the years the breed became a popular contender for main show awards and for long periods it attracted some of the most competitive breeders, first in large fowl and then, when they were established, in a bantam version. Black, whites and blue have been popular colours, but contemporary bantams are now being seen in more complex colourways that look like examples of multi-coloured Minorca-type fowl as seen in early Dutch paintings. The large Minorca was never in the top rank of commercial layers, but it always had a following among urban domestic poultry keepers where it was found they withstood confinement well. The breed used to be a favourite of domestic keepers, with at least one known strain laying more than 230 eggs per year.

▶ *A large Black Minorca male with the breed's classic wedge-shape body, bold, upright comb and clean, almond-shaped lobe.*

▲ *This bantam Minorca has attractive almond-shaped lobes.*

ESSENTIAL CHARACTERISTICS

Size: Large male 4.3kg/9lbs. Large female 3.4kg/7½lbs. Bantam male 963g/34oz. Bantam female 850g/30oz.
Varieties: Black, Blue, Buff, White.
Temperament: Flighty.
Environment: The large comb makes the breed unsuitable for cold areas.
Egg yield: Large white eggs.

Dandarawi

Although indigenous to the eastern Mediterranean, the tiny Dandarawi also belongs to the European landrace group of breeds. The term landrace describes types of poultry (as well as other birds and animals) that are not necessarily formalized breeds, but represent a local population of poultry that has developed in response to the environment in which it lives. Its essential characteristics and visual appearance are determined by its ability to survive in its circumstances, whether the result of modifications or human intervention. Populations can be landraces and they can also be standardized breeds.

▼ *This Grey male has a rudimentary buttercup comb and five toes.*

The Dandarawi is a small bird. By nature it is flighty, and does not take kindly to close confinement. It is therefore not a bird to choose as a family pet.

Males are black with a white hackle and saddle and some white on the wings and body. Females are wheaten with white, grey or reddish-brown markings. This is an auto-sexing breed in which the young can be sexed upon hatching. Typically, the female can be identified by a black blotch on the head or stripes on the back; the male has no such blotches. The crest of this breed points backwards. These examples of a non-standardized population rather than a breed are included to illustrate the sort of local or village populations that exist throughout much of Northern Africa and the Near East. These populations are variously described as Berber and

◄ *It is just possible to see a small lark's crest on this four-toed female. She looks to be duck-footed, a fault that appears occasionally in nearly every standardized breed.*

Bedouin fowl by those studying the origin of Mediterranean poultry breeds. They have rudimentary buttercup combs, which may provide some indication of the type of genetic material that went into making the Sicilian breeds which share the same feature. Dandarawis mature at about six months of age and may lay up to 150 eggs per year. Small groups of this sort of fowl occasionally turn up at shows, entered by enthusiasts. The birds are of interest to other enthusiasts, but will usually bemuse those exhibitors used to breeding fowl to exacting standards. In fact, this sort of fowl is best seen as an example of a population that evolved through environmental pressures. Conscious selection would soon diminish the population's overall genetic integrity.

ESSENTIAL CHARACTERISTICS
Size: Small.
Varieties: Not standardized.
Temperament: Flighty. Ability to survive on very low nutritional plain.
Environment: Free range. Naturally harsh and arid.
Eggs yield: Will improve given good nutrition.

White-faced Black Spanish

Much of the fowl once indigenous to the Iberian Peninsula seems to have been selected and standardized with large white lobes and a wedge-shaped body. It seems likely that fowl of this rudimentary form, later standardized in its homeland as Castilian, found their way into the Low Countries in the 16th century at the time of the occupation of the Spanish Netherlands. The White-faced Black Spanish is known to be an ancient breed, close in its breeding to the Minorca and Castilian breeds.

Huguenot weavers are said to have kept Spanish fowl in Spitalfields, London, in the late 1600s, and the breed was reported as popular in Bristol by the 1850s. While many of its breeders would have been obsessive in their selection for size and perfect texture of the white face, it seems likely that this one breed has always been in the hands of relatively small groups of dedicated fanciers. At several times in its long history it has come close to extinction.

The urge to perfect or even exaggerate some natural feature of a breed seems to be deeply embedded in some cultural groups. White-faced Black Spanish fowl have bigger white lobes than those of other fowl, and in some cases this white area extends to that part of the face that is normally red. It is likely that these are all

▼ A large Spanish showing the characteristic white face that is expected to extend above the eyebrows almost to the base of the comb.

features that have been exaggerated by early breeders. Because of its coloration the breed has also been known as the Clown-faced Spanish. Just what prompted an individual to select and develop this feature, until it extended over most and eventually all of the face, we will never know. Although the over-developed white earlobes are the most distinguishing feature of this breed, it also has black plumage and a single comb. It has four toes and lacks a crest.

As recently as 1970, the breed had diminished in numbers to just a handful in Germany. An international effort to save it was successful, and saw the breed recover to a point where breeders could spend time selecting and perpetuating its huge white face. There are still years when the breed all but disappears from the show scene, only for several excellent examples to return the following year.

Keeping its white face free from disfiguring blisters may mean that it needs protection from rain and wind. However, since generations of the breed have been kept in close confinement, it is more content than

most other breeds in an urban environment, provided it is well cared for. Few breeds have had to wait so long for a bantam version, but its creation in 1980 means that those with limited space can now keep the smaller version of the breed. Both large and bantam versions lay quantities of snow-white eggs.

The breed does not go broody, and chicks are born with white facial feathering. The white face takes up to one year to develop fully, becoming more opaque as the bird ages. Selecting this unique, completely white face to exhibition-standard perfection means that breeders often only retain relatively small numbers of both males and females for future breeding. The downside of this search for perfection is that it could lead to a diminished gene pool, yet at the same time major show awards have been won by breeders who keep rather more hens for their utility, rather than exhibition, qualities.

ESSENTIAL CHARACTERISTICS

Size: Male 3.6kg/8lbs.
Female 3.2kg/7lbs.
Varieties: One only.
Temperament: Noisy, alert, friendly.
Environment: Copes well with confinement.
Egg yield: 150–220 large white eggs per year.

Blue Andalusian

The black poultry breeds native to Spain occasionally produce slate- or grey-coloured offspring. It is likely that the origins of the Blue Andalusian breed can be found in this lighter-coloured poultry. While it was once fashionable to claim foreign or exotic ancestry for every new poultry breed, it is likely that this breed was made and named in Britain. The Blue Dun Game poultry breed may have been included in some strains of its ancestry; however, most would have been "formula bred" by back-crossing grey sports of black Spanish breeds that would have been related on both sides. After a few generations, the colour of 25 per cent of the offspring would be a very pale grey splashed with the odd blue feather. When these were crossed with related black birds, all of the offspring had blue feathers with black edges. When these blue fowl were mated together they regularly produced 50 per cent of their offspring with "blue" plumage, 25 per cent with "splashed" feathers and the rest, black. The formula was

▶ The male Blue Andalusian has an upright comb and laced feathers on a ground colour that is difficult to perfect.

based on the work of Gregor Mendel. Blue Andalusians were successfully bred to this formula in the 1850s and 1860s. The early examples were vaguely blue-grey with rudimentary lacing, but over the years dedicated breeders have managed to produce birds with crisp blue-black lacing around a ground that is close to giving the illusion of being blue. It is this degree of breeding difficulty and the urge to create something beautiful that attracts many breeders. Yet, over the years, this breed has relied on three or four dedicated breeders to stop the breed from disappearing. A bantam version could help create interest in the breed, but keeping enough Blue Andalusians to provide the necessary gene pool is a formidable task. By nature the Blue Andalusian is known as a noisy bird. It has a graceful gait and prefers a free-range environment to close confinement. In spite of their largely Spanish ancestry, they can be good winter layers.

◀ Many of the exhibition strains of the Mediterranean breeds are heavier than the original strains. The Andalusian, possibly due to an infusion of English or Spanish game in its ancestry, often carries more meat than Leghorns. Interestingly, black sports look like the native Spanish Castilian breed. This is a hen.

ESSENTIAL CHARACTERISTICS

Size: Large male 3.2–3.6kg/7–8lbs.
Large female: 2.3–2.7kg/5–7lbs.
Bantam male 680–793g/24–28oz.
Bantam female 566–680g/20–24oz.
Varieties: Blue.
Temperament: Active and agile. Likes foraging.
Environment: Free range.
Egg yield: 166–190 white eggs per year.

Sicilian Buttercup

The Sicilian Buttercup may represent a link between fowl native to the eastern Mediterranean, as its name suggests, and other Italian breeds that have played an important role in commercial poultry breeding, such as the Leghorns.

The first thing one notices about the breed is its unusual, and sometimes spectacular, comb, which is a feature common to the few other breeds that have a shared ancestry. On the body, the broken black barring or pencilling on a bay to rich golden ground is almost identical to that of many Egyptian fowl, such as the Fayoumi breed. The female is noticeably lighter in colour than the male, with a soft buff ground, while the male has a deep, rich orange ground. While Fayoumis usually have grey or slate legs, most of the Sicilian Buttercups have green or olive legs. This suggests that there has been a certain amount of interbreeding with other yellow-legged Italian breeds such as the Leghorns, which in turn, have had earlier contact with breeds imported from the Far East.

The Sicilian Buttercup first came to popularity during World War I when, for the first time, poultry was rationed as a food source. However, eggs were only rationed by price, and people wanting an abundant supply of eggs prior to the war were willing to pay premium prices for them. The Sicilian Buttercup breed was known for its vigour at egg production and as a result, it became a popular breed in wartime. For a while there was sufficient interest and demand for a Sicilian Buttercup breed club to be formed. During this time, a second Buttercup breed known as the mahogany-coloured Sicilian Flower emerged.

ESSENTIAL CHARACTERISTICS
Size: Male 737g/26oz.
Female 623g/22oz.
Varieties: One colour.
Temperament: Wild, flighty.
A good flier.
Environment: Free range.
Egg yield: 200 small white or tinted eggs per year.

With the return to peacetime, and the plentiful availability of large eggs laid by rapidly improving breeds such as the white egg-laying Leghorns, interest in the Sicilian Buttercup breed waned and there has been insufficient interest to maintain a worthwhile gene pool. Today the breed is so rare that one cannot rely on finding really good examples among the rare breed classes, even at the larger poultry shows. Aside from the breed's positive utility qualities, it does have an

▲ *The exhibition comb is expected to be as round and bowl-shaped as possible, with an evenly castellated rim.*

extremely flighty nature, a characteristic that many breeders will tolerate in order to keep a breed with an almost unique comb. The Sicilian Buttercup is a sufficiently good layer to breed the sort of numbers needed to select future breeding stock with very good combs.

While bantams with buttercup combs have been seen, these have never been classed as miniature Sicilian Buttercups.

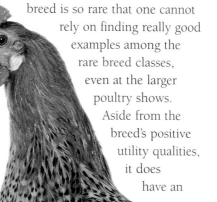

◄ *The hen's lightweight frame, rather elongated body shape and broken black barring or pencilling suggests a close relationship between this Sicilian breed and poultry breeds of the Eastern Mediterranean, such as the Turkish breeds as well as their Northern European Hamburg-type relatives.*

Leghorn

This is an old breed and one of the most well known. It is thought that the breed became known as the Leghorn in America because the first shipment that arrived there in 1853 came from the Italian port of Livorno, pronounced "leghorn" in English. One cuckoo-coloured and four white hens, formed the initial shipment, together with one rooster.

Whether as egg-laying fowl or exhibition birds, nearly all Leghorn poultry selection took place outside of Italy, notably in America, the UK and Denmark, where all colours of the breed have been developed to the most exacting standards. The majority of fowl varieties seen today are brown, white and intermediate colours with, as a rule, yellow legs, although many of the black variety have slate or black coloured limbs. The breed is exhibited with a wide range of differing standard colours.

The breed displays a natural reluctance to broodiness, a characteristic that the commercial egg industry has capitalized upon. Leghorns have historically been highly regarded for their egg-laying capacity. White hens can produce on average 250 pure white eggs per year, and as a result the breed's genes have been used extensively in commercial egg-laying hybrids.

▶ *A proud upright is a classic feature of leghorn males.*

▶ *This Brown bantam hen has the characteristic large floppy comb. Breeding birds to an exacting physical standard for exhibition means that the bird's egg-laying capacity is reduced.*

This selection of the breed for its egg-laying capacity saw the Whites reach a point in the 1930s where they were responsible for producing most of the world's commercially produced white eggs. White Leghorn-based hybrids convert poultry feed into eggs more efficiently than any other pure breed, which is particularly impressive when the breed's light body weight is taken into account. Breeding for the splendid perfectly modelled upright male combs, and female combs that flop gracefully to one side, may have detracted from the peak of utility performance. Many of the large fowl versions bred for showing may be bigger than they need be to fulfil a utility function, but this size has enabled the creation of the bantam versions that are, in most colours, exact counterparts of their larger versions.

Selection to keep their size small in some bantam strains means that some families of Black Leghorn bantams, for instance, lay almost as well as commercial hybrid versions. All have an active disposition, but few are as flighty as many other light breeds.

As one of the great international breeds, Leghorns throughout the world vary more than any other poultry breed. Exhibition strains of the breed have built on the statuesque stance of the first White Leghorns imported to America. American breeders have favoured the marginally more horizontal body and gracefully sweeping tail carriage. German breeders, who only use the title Leghorn for the white variety, have bred a more horizontal outline. Other varieties are known in Germany as "Italians", and have an exaggerated horizontal tail and carriage.

ESSENTIAL CHARACTERISTICS

Size: Male 2.4–2.8kg/5½–6½lbs. Female 2–2.5kg/4½–5½lbs.
Varieties: Black, Black-tailed Red, Buff, Columbian, Dark Brown, Light Brown, Red, White.
Temperament: Flighty, noisy.
Environment: Prefers free range but utility strains have adapted to confinement.
Egg yield: Strains of utility Whites can lay up to 300 white eggs per year. Exhibition birds may lay fewer.

Ancona

The Ancona originates from the town of the same name in Italy, and was taken to England in the mid-19th century. It has similarities to the Leghorn and was initially known as a Mottled Leghorn. It is this mottling, the distinctive white ticking on black plumage, which is the distinguishing feature of this breed. As such, the breed standard for feather marking is exacting. Breeding to meet the precise feather patterns required may have saved some strains from excessive head point development. An alterative rose comb is more often seen in the Ancona breed than in Leghorns. The breed has a large, red single comb which may flop in females, and long red wattles. The legs are deep yellow and the build is slender.

By nature this bird is an active forager, easy to tame, and with a quick temperament. The earlier utility flocks of large Anconas may no longer exist, but some strains of the bantam version are capable of laying quantities of surprisingly large eggs.

▼ *The black feathers with white tips give a mottled appearance, and overall the plumage has an attractive greenish tinge. This is a bantam male.*

ESSENTIAL CHARACTERISTICS

Size: Male 2.7–3kg/6–7lbs.
Female 2.25–2.5kg/5–5½lbs.
Varieties: Mottled.
Temperament: Active, alert.
Environment: Free range but adapts to confinement.
Egg yield: More than 160 white eggs per year.

Fayoumi

The Fayoumi is an ancient breed that originated in Egypt. It is a small, lightweight bird: males weigh just 2kg/4½lbs and females 1.5kg/3½lbs. It is now kept for ornamental purposes, but it was traditionally kept for its egg-laying capacity, producing small, off-white eggs. The breed is characterized by a forward-thrusting neck and chest and an upright tail. Its flesh is slate blue, and the wattles, comb and earlobes are all red. This breed can move very quickly, is good at flying and dislikes captivity. If caught, this bird emits a noise similar to the cry of guinea fowl. Fayoumis are a hardy breed, being tolerant of extreme heat in keeping with their North African origins. As lightweight, good foragers they can have very low maintenance requirements, and when kept in well-ventilated conditions have much resistance to disease. These birds are quick to mature.

ESSENTIAL CHARACTERISTICS

Size: Male 2kg/4½lbs.
Female 1.5kg/3½lbs.
Varieties: Gold-spangled, Silver-spangled.
Temperament: Dislikes handling and is extremely flighty.
Environment: Free-range, even living in trees.
Eggs: Egg numbers can be high but are small in size and white or tinted.

◀ *Like many Fayoumis, this male has only a rudimentary barring on a golden ground colour.*

ASIAN SOFT-FEATHER FOUNDATION BREEDS

Many of today's domestic fowl are descended from Asian domesticated breeds, which historically were kept in densely populated areas, often in coastal margins. These are all massive birds, true to their original status, but it is likely that in appearance they have little in common with the first birds imported from Asia. These breeds have been selectively bred for their size, and have been used extensively to provide genetic material for other breeds.

Cochins and Brahmas have a significant amount of feathers that adds bulk to the visual appearance of these breeds. For all their size, however, they are gentle giants and make lovable pets for children as well as fabulous garden companions. Javas and Langshans have an equally attractive nature and like to range free. All will need large-scale housing and plenty of room in which to range. Due to their size they eat a large amount.

Cochin

The Cochin breed was responsible for creating a craze that became known as "poultry mania". Large fowl that were later reported as coming from Cochinchina, were probably the result of many different imports from various Chinese and Asian ports. The birds first reported in newspapers as being brought from Cochinchina, by Sir Edward Belcher, the scientist and explorer, were given to Queen Victoria in 1843. However, an analysis of his log on *HMS Sulphur* shows one reference to his loading fowl aboard in what is now Northern Borneo. Originally known as Shanghais, the five pullets and two roosters that he collected later became known as Cochin-Chinas. Later still, the name was abbreviated to Cochin. Cochin is a port in India and it is likely that the birds were named for this port on their

ESSENTIAL CHARACTERISTICS

Size: Male 5kg/11lbs. Female 3.9kg/8½lbs. For bantams *see Pekin*.
Varieties: Black, Blue, Buff, Gold-laced, Partridge, Silver-laced, Splash, White.
Temperament: One of the friendliest poultry breeds. Quiet.
Environment: Perfect for the home garden.
Egg yield: 120 dark brown eggs per year.

▶ *This large Black Cochin hen has feathered feet, a characteristic feature of the breed, and a propensity for extremely fluffy, almost silky, under-fluff.*

journey to the western world. The fowl Queen Victoria was given were reportedly good layers; this was borne out when a hen exhibited by the monarch in Dublin was reported to have laid 94 eggs in 103 days. The breed was also rumoured to lay two eggs per day, and because of its enormous size, was considered a good breed to rear for the table. Interest in the Cochin was heightened, and soon entrepreneurs were scouring the Far East for similar fowl. Northern Chinese ports were open to trade with the West, and fast tea clippers carried live birds from East to West. Unlike known European breeds, many of these birds had feathered legs. A second batch of birds was shipped from China in 1847. Although these were sold as Cochin, they bore little resemblance to the birds delivered a few years earlier. It is unlikely that the birds that were originally shipped from the East share any physical resemblance with the Cochin breed that we know today.

These early Shanghais were used to improve many local populations of European fowl, and the results became the basis of utility, brown egg-laying breeds such as Welsummer, Marans and Barnevelder. Desirable attributes of the Shanghai included foot feathers, and great numbers of fluffier feathers made the birds look enormous. These features were selected and bred into local poultry as well as altering the appearance of the Cochin breed.

The fact that such breeding tended to reduce the birds' natural utility was never a consideration for breeders of exhibition birds. The early boom in the breed had seen stock widely distributed, and if anyone wanted table fowl, a Cochin crossed with a Dorking-type fowl provided plenty of meat. Similarly, a pullet bred from a Cochin and a farmyard hen laid as well as any other bird. The craze for such poultry was short-lived when it was realized that although the birds appear large, it was merely the feathers that made them look bigger.

As an exhibition breed, Cochins had, by the 1890s, been selected to assume a shape, form and degree of feather that would be recognized as standard today. The breed now had the ability to produce attractive feather and fluff in a new range of feather patterns and colours. In China, the feathers were used for padding and insulation. Many of the fowl shipped from Shanghai may have been Black, but other imports

▶ *The Cochin breed was originally kept for its dark brown eggs and for its meat. This is a large Buff-coloured male.*

▲ *Cochins make adorable pets. This is a Blue-partridge rooster.*

included a natural Buff that had not been seen in the West previously; initially it was described as Cinnamon. This colourway was eventually refined to become a clear even buff that we now see in other breeds. The Partridge colour, with its intricately marked female form, was very different to the black-reds of either jungle fowl or native game breeds. This brought new complexities to exhibition varieties in other breeds. Whites that had previously been seen as weak when occurring as

sports in other breeds now became a feature to select, and a degree of whiteness is still a major show challenge in many breeds. Breeders introduced the gene of "Columbian restriction". Selection for overly soft feathers, body fluff and foot feather, which were seen as exhibition points soon detracted from the breed's original utility properties. For most of its existence, the Cochin has been anything but a commercial or even utility breed.

Today, the attributes that made the breed popular as a utility bird are those that make it an adorable, fluffy pet. Cochins have a gentle and docile personality, despite their large size. The bird has an aristocratic nature, and is identifiable by its size and the large cushion of feathers on its back. The male bird has a deep crow. When not bred for show, it is best seen as a delightfully placid ornamental companion in the garden or small paddock.

Brahma

A true giant among poultry breeds, the Brahma is thought to have the ancient giant jungle fowl, *Gallus gigantus*, in its ancestry. The breed originates from the Brahmaputra region of India, where one probable ancestor was known as Chittagong. It was first shipped to America in the middle of the 19th century, where it was given the name Brahma Pootra and eventually the name of Brahma. It is thought to have been selectively bred with a Malay-type breed in the USA, and the resulting offspring imported to Britain almost a decade later. Such breeding may have been responsible for producing a fowl with a pea or walnut comb as characteristic of the hard-feathered breeds of Asia. The comb soon became a standard point for show specimens.

In his book of 1855, entitled *Hen Fever*, George Burnham, an American breeder and entrepreneur, recalls how he selected a few large grey birds from a consignment of several hundred fowl of all grades and

▼ *A large Dark male, one of the original colour varieties that played an important part in creating the Sussex breed.*

▶ *A large Gold Brahma, taller and far more statuesque than the related Cochin.*

proportions imported from Shanghai, China. He crossed these with his existing fowl, which were likely to be heavy-boned breeds (Cochins) from different locations. "I selected from my best Shanghai chickens nine beautiful birds. They were placed in a very handsome walnut-framed case and having duly landed [from China]

▼ *Brahma bantams can be larger than other bantam breeds but are often similar to their large counterparts.*

were forthwith shipped across the big pond addressed as follows: 'To Her Majesty Victoria Queen of England'." The new birds, housed in London Zoo, fed the poultry mania sweeping the country.

Some of the original male American imports weighed up to 6.4kg/14lbs. The original grey colour of the imported birds was quickly distilled into Light and Dark varieties in the UK. The genetically created Silver-Columbian pattern provided a significant gene pool for the Light Sussex breed. The Dark variety later found its way into varieties of breeds that claim a wholly European ancestry. Over the years, Brahma breeders have been able to build on the original complex patterns to make

▲ *A Buff-Columbian female bantam, in which the black coloration is restricted to the hackle, tail and closed wing feathers.*

▲ *Brahmas have a distinct tail that is held aloft. This is a Gold Partridge hen.*

▲ *The Light Brahma variety has a restricted black feather pattern.*

and standardize new colours and feather patterns. Today there are several varieties, including the original Light, Dark, Gold, Black, White, Blue, Buff-Columbian and Blue-Columbian.

As Brahmas are so large, birds can take two years to develop to full size, and so were never a commercial breed. Some strains retained enough of their original utility properties – the bird was at first considered a dual-purpose breed – to be used to help make some of the first broiler crosses in the 1930s. Once a prolific egg layer, the additional colours that have been bred into the Brahma were at the expense of its egg-laying ability.

Nowadays the breed is considered a pet or is kept for exhibition, since its utility features are no longer so

important. The giant size, feathered feet and calm disposition make this an ideal garden breed. It is essential to train this breed from a young age however, since its large and weighty build may make it difficult to handle when it is mature. For those with

limited space, the bantam versions, which were popular in the 1890s, can still be enjoyed. It is a most interesting, beautiful, challenging and hardy breed. Despite the size of this large bird, care needs to be taken that it is not bullied by other breeds, including much smaller birds that can sometimes take advantage of the breed's docile temperament.

ESSENTIAL CHARACTERISTICS

Size: Large male 5.4kg/12lbs. Bantam male 1kg/38oz. Large female 4.1kg/9lbs. Bantam female 907g/32oz.
Varieties: Black, Blue-Columbian, Buff-Columbian, Dark, Gold, Light, White.
Temperament: Calm and friendly. They make good pets.
Environment: Garden.
Egg yield: 120 brown eggs per year.

▶ *A large Dark Brahma hen with a coloration that in other breeds would be known as Silver-pencilled or Silver-partridge.*

Langshan

Most of the birds that early poultry keepers called Cochins (intending to capitalize on the early popularity of those delivered to Queen Victoria), were shipped through the northern ports of China. Many of the black examples came from what we now know as the port of Shanghai. Just how many of these black fowl were similar to the later importations of Langshans, or if any of these laid eggs with the characteristic plum-coloured bloom that become associated with the breed, is impossible to say. Since many of the birds that were originally imported from Asia were selected and bred from to produce new breeds for exhibition, each with specific visual characteristics, the utility value of many of these birds suffered. To find fowl with the original utility potential meant importing more of the ancestral fowl from the point of origin. However, finding the exact location would have met with some confusion. Shanghai, Lan-Chau, Yangtse or Langshan were all either native or European names for the same general region of China.

It would have been from this general region that Major Croad, in 1872, made one of the last (and arguably one of the least adulterated) importations; the breed is still believed to exist in that region of China. Major Croad concentrated on breeding the utility properties of his imported birds, and he eventually linked genes contained within his strain to produce what we now recognize as the Croad Langshan type.

The breed was kept initially as a utility breed, and was developed for both its egg-laying capacity and its production of very dark brown eggs. Asiatic breeds lay browner eggs than those otherwise seen in Europe. Later breeders sought to emphasize the differences between most of the Asiatic breeds.

The bird has a broad outline and a large, upright body shape. It stands tall on its legs and carries a single upright comb. The black colour of the plumage has an almost bottle-green sheen. It is now thought that the spectacularly U-shaped outline of the breed and the stiff upright tail carriage of the bird promotes the production of dark brown eggs – a fascinating genetic link. The upright tail feature is very different from the cushion tail associated with the Cochin standard.

Another interesting facet of this breed is the plum-coloured "bloom" that often overlays the surface of the brown egg, which always attracts attention at exhibitions.

Over the years there has been more than one competing Langshan breed, but the original Croad is still kept in sufficient numbers in both large and bantam forms to ensure the survival of the breed. The Langshan has an inquisitive but docile nature and can cope with confinement as well as free-range situations, making it an ideal pet. The breed is known to produce good mothers.

The original Croad Langshans were exhibited at the same time as the early Black Orpington, which was

▶ *A Large Croad male with characteristic short back and lightly feathered legs.*

ESSENTIAL CHARACTERISTICS
Size: Male 4.3kg/9½lbs. Female 3.4kg/7½lbs.
Varieties: Black, Blue, White.
Temperament: Friendly, docile.
Environment: The Croad variety can thrive in a wide variety of environments, but the German bantams prefer a more protected lifestyle.
Egg yield: The Croad can lay 140–180 brown eggs with a plum bloom per year, with some strains being noted winter layers. Exhibition-bred bantams will not lay as many or such attractively coloured eggs.

said to have some Cochin stock in its ancestry, and which looked similar to the Langshan at that time. In order to make the two breeds look different, some exhibitors bred taller birds. Those breeders of the original strain continued to promote what they saw as the correct type and formed the Croad Langshan club. Breeders of the tall birds named their birds Modern Langshans to distinguish their exhibition breed. Some examples had sloping backs and tail carriage and very scanty leg feathering, with certain authorities insisting that some had difficulty in standing. While the traditional Croads prospered, the Modern Langshans were scarce from 1920, and probably died out around 1960. However, prior to this date the breed was taken to Germany, and

▼ *Exhibition forms like this German Langshan have been exaggerated to a point where they look nothing like the original.*

from this stock the German Langshan was bred. From German Langshan stock it has been possible to recreate the original bird.

German Langshans, while tall, with clean legs, developed the same U-back and tail carriage as the Croad Langshans. The large German version is not quite the size of the Croad Langshan. A few large fowl have been seen at shows, but the bantams were soon established as one of the most popular and successful at exhibitions. The breed has been selected for utility and for exhibition, a process of selection that can lead to exaggeration in feather characteristics and perceptions of breed type. However, the tail of this breed is more rounded than that of the original birds.

▶ *A Black German Langshan bantam male, with typical upright tail.*

▼ *A Blue German Langshan bantam male with standardized clean legs.*

Croad Langshans are a heavy breed, but carry their weight gracefully. The head of the Langshan is small in comparison to the rest of its body. The crest is carried upright in male and female birds. The breed is active but docile and tolerant by nature and does not mind being handled, making it a good breed to keep if young children are around. The hens make excellent mothers. The breed is long lived, with some birds living up to 10 years. Egg-laying capacity is known to diminish rapidly after middle age, however.

The Langshan breed development mirrors much else that has happened in exhibition-based poultry selection. The original Langshan from Shanghai would have been very similar to those earlier useful fowl that helped to make what became known as Cochins. Those remaining with Major Croad's family retained a similar shape to the useful Croad still found today.

Java

This breed was imported to the USA from the East Indies in 1883 and was developed and refined there. Its Asian ancestry is unknown. The breed is little known on the world stage and is now quite rare in the USA. Nevertheless, the Java is one of the foundation breeds of modern poultry and is one of the oldest American breeds along with Plymouth Rocks and Jersey Giants, to both of which, it is known to have contributed genes. The Java belongs within the family of heavy, soft-feather breeds that had a wide distribution throughout much of Asia. Unlike those birds exported from China, the original fowl did not have feathered legs or feet.

In appearance, it is a large sturdy bird, with a long, low back and a plump chest. It is one of the largest poultry breeds. It has small ear lobes and a medium size comb, which starts quite low on the face suggesting that it has an ancestral link to breeds with a

ESSENTIAL CHARACTERISTICS

Size: Male 4.3kg/9½lbs. Female 2.9–3.4kg/6½–7½lbs. No bantam.
Varieties: Black, Mottled, White.
Temperament: Docile.
Environment: Free range. Prefers a small flock.
Egg yield: Reasonable numbers of mid to dark brown eggs.

pea comb. Black and mottled varieties are available, though white was developed and later abandoned due to its similarity to another popular breed. White chicks do appear occasionally, and dedicated breeders, are keen to re-establish the variety. The black variety has beetle-green plumage, black legs and beak, while the mottled and white varieties have a yellow beak and legs and white flesh.

▲ *The breed was once known as Black Java, suggesting that black was considered the typical breed colour. It is still a popular choice of variety.*

This is a dual-purpose bird that is economical to keep if allowed a free-range lifestyle. The hens go broody and are known to make good mothers. Like most of the heavy soft feathered foundation breeds imported into Europe and the USA, the nature of the Java is that of a very docile bird. Cochins may also have originated in the same general area as the Java breed.

▶ *Most black varieties of fowl produce an occasional splashed or mottled sport.*

◀ *The Java hen is considerably smaller than the male of the breed.*

GAME BIRDS AND ASIAN HARD-FEATHER

The hard-feathered breeds of India, Malaysia and, to a lesser extent, Thailand, have had a significant influence on modern poultry development. These hard-feathered breeds have a high stature, but are devoid of fluff in their plumage and are poor and seasonal layers. They differ in practically every way from those breeds native to Northern Europe, yet intriguingly, they also differ from the jungle fowl that inhabit the same part of Asia. This harder, fluff-free feathering could have evolved as domestication took the descendants of jungle fowl to hotter lands, far south of their native regions. In the case of the Aseel, the feathering developed into a form of body armour. Indian or Cornish game were never used for cockfighting, but with the Aseel as one of their foundation breeds, they went on to develop a high muscle-to-offal ratio. Indian Game form part of the ancestry of the modern broiler fowl; the breed inherited much of its bulk from Malay-type fowl from Eastern India. Aseel is the ancestor of much of the world's economically important poultry stock. Breeders of fighting birds would have described all their fowl as "game". However, other often closely related strains were soon being described as separate breeds, named after their breeder or by their colour.

OLD ENGLISH GAME

Old English Game fowl are the descendants of the original pit-fighting birds bred by the nobility. They are highly prized, and extremely desirable, and exhibition varieties can command huge prices. When cockfighting was outlawed in England in the mid-19th century, the fighting birds were bred instead for exhibition purposes. At the time the country was gripped by a craze for the newly imported giant, soft-feathered Asiatic poultry, and the size of these imports had an influence upon the breeding of Old English Game. Soon exhibition judges were selecting the biggest and tallest birds for championship prizes.

Fowl that had been selected for the ability to survive the rigours of the cockpit evolved into strains. The strains developed according to type, size and colour, depending upon the characteristics that each breeder sought to emphasize. Each was influenced by his or her interpretation of the breed standard. The exhibition standard for the breed used a form of wording that reflected the breed's history as pit game, and this remained the only written standard used for this breed for almost a hundred years. In the 1930s the Old English Game Club split, and today, two distinct types of game are bred: Oxford and Carlisle.

THE SPLIT WITHIN THE OXFORD GAME CLUB

Writing about events that led to the formation of the Oxford Game Club, Herbert Atkinson, one of the original members, wrote.

"I fancy the general public believed the true Game Fowl to be a thing of the past …. Here let me say that it was solely among cock-fighters that he did still exist. Many of these had the same breed for generations. They were not led away by showing or fashion or moneymaking. Their requirements were purity of blood and courage, activity, strength and sound-ness of constitution." The adoption of an agreed standard in the 1880s for the Old English Game breed would have had to reconcile various concepts about differing strains. It would not have been the wording of that standard, but its interpretation, that led to splits within the game fraternity. The rift happened in the 1930s, and at that point separate Oxford and Carlisle Game Clubs were formed. Oxford birds are considered to be true to their cockfighting roots, whereas the Carlisle type are influenced by the giant Asian poultry being bred for exhibition, and are of a larger build.

◀ A Mealy-breasted Mealy-grey Oxford male.

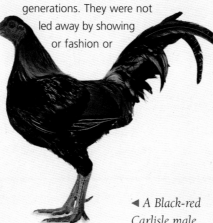

◀ A Black-red Carlisle male.

Oxford Game

The game fowl categorized as Oxford type was the product of centuries of selection from local populations of game strains. Today's fowl are similar in bone structure and feathering to their ancestors, and are only slightly larger. Any increase in muscle development, weight and bone structure would have been kept in check by the bird's inherent need for speed and agility. The extent to which breeders incorporated the different hard-feathered Asiatic breeds, particularly Aseel, into their breeding programs varied considerably. Much depended upon each breeder's requirement to improve his stock in order to select the best of the breed, with the intention of matching the written standard. Oxford Old English Game are feathery. The small head is furnished with a single comb, small lobes and wattles. These game birds have large eyes and, appropriately for a fighting bird, a strong and powerful beak, a lean and muscular build, with short and strong legs, powerful wings that slope down to the ground at an angle and a sloping back. These latter two characteristics help to distinguish the Oxford type from the Carlisle type. (An increasing number of shows put on classes that accommodate both Oxford and Carlisle types, and this helpfully allows the layman to see the differences between the two.) Unlike Carlisle Game, the Oxford fraternity stick rigidly to the upper weight limits of their standard for exhibition purposes.

◄ A Grey male.

Oxford Game are more numerous than Carlisle Game. The Oxford standard lists 32 colour combinations that often reflect a strain's earlier connection with a trade or district. Game men of all persuasions quote the old truism: "There is no such thing as a bad colour in a good game fowl".

The term "hard feather" is usually used to describe the tight feathering and glossy, watertight characteristics of the plumage. In relation to the Oxford Game birds, it seems to have been used to describe the quality of the bird. Soft would have suggested an unfit or out-of-condition bird. It is customary to dub male game birds, which means removing part of the comb and wattles, originally so that they did not get damaged when the bird was fighting. Continuation of this tradition is likely to remain an issue.

◄ A Partridge hen.

▶ A Pile male.

ESSENTIAL CHARACTERISTICS

Size: Male 1.7–2.6kg/3lbs 12oz–5lbs 12oz. Female 1.8–2.2kg/4–5lbs.

Varieties: Black, Black-breasted Birchen-duckwing, Black-breasted Dark Grey, Black-breasted Dark Red, Black-breasted Red, Black-breasted Silver-duckwing, Black-breasted Yellow-duckwing, Brown-breasted Brown-red, Brown-breasted Yellow-birchen, Clear Mealy-breasted Mealy-grey, Cuckoo, Dun-breasted Blue Dun, Furness Brassy Back and Polecat, Hennie, Ginger-breasted Ginger-red, Muff and Tassel, Pile, Spangled, Streaky-breasted Light Red, Streaky-breasted Orange-red, Streaky-breasted Red-dun, Yellow-silver and Honey-dun, White.

Temperament: Agile and able to fly over most fences.

Environment: Free range.

Egg yield: More than 100, with some strains capable of laying far more medium-tinted eggs per year.

Carlisle Game

All Old English Game are light and flighty birds with a wild and alert quality. Most have powerful wings and can fly, and this should be taken into account when housing the breed. These are birds with a strong, athletic and muscular build. They have been bred for the pit and have an inherent instinct to fight. Male birds will fight to the death and should never be penned together. The hens are known to make good mothers.

The poultry fanciers who initially bred Carlisle game had collectively selected their birds to fulfil the exhibition requirements for larger fowl. As a result, Carlisle Game birds are larger birds than the Oxford type. Both versions incorporate Aseel blood in their genes. Many of the birds in Carlisle classes at exhibition exceed their breed's standard weight considerably. They also differ from Oxford Game in their posture, as Carlisle Game birds hold their back and wings horizontally. The birds have a deep breast, and because of this were considered meat birds, with a strong, robust flavour. Carlisle game are standardized in fewer colours than Oxfords, although there are 13 accepted varieties, more than for many other pure poultry breeds.

▼ *A Black-red male.*

▼ *In order to assess their fowl, Carlisle Game fanciers hold the bird so that it faces towards their own chests.*

▼ *A Brown-red male.*

▼ *To assess overall balance and agility, judges and exhibitors of Oxford Game birds instinctively hold their lighter birds so that they face away from their own chests.*

ESSENTIAL CHARACTERISTICS

Size: Male 2.9kg/6½lbs. Female 2.4kg/5½lbs.

Varieties: Birchen or Grey, Black-red (partridge), Black-red (wheaten), Blue-red, Brown-red, Crele, Cuckoo, Golden-duckwing, Pile, Silver-duckwing, Spangle, Blue-tailed wheaten, Self-white.

Temperament: Vigorous but not as agile as Oxford game.

Environment: Free range. Can fly.

Egg yield: Up to 130 smallish white or tinted eggs per year.

Aseel/Asil

The Aseel is an ancient breed originating in India. It is mentioned in the Codes of Manu, the ancient Indian legal texts, and is considered to be the oldest documented breed of poultry in the world. The name Aseel translates to mean foundation, original, regal or noble.

Merchant adventurers trading with India were attracted to the fearless fighting qualities of this breed, and it is they who were responsible for its export to other parts of the world in the early 1800s. Aseel-type fowl found their way into the cockpits of Britain. However, this arrival seems to have made little impact on English game breeds, which appear to have remained feathery and generally more like red jungle fowl than the Aseel. Nevertheless, the Aseel has had an influence on the development of some strains of Old English Game.

The breed almost certainly played a formative role in the development of the Indian/Cornish Game breed, which has added much to the shape and conformation of the broiler fowl that is used to supply much of the world with poultry meat.

In appearance the Aseel has strong and heavy muscles. It has an upright stance with a thick and muscular curved neck and a strong beak. It

▼ Aseels are intelligent, aristocratic and pugnacious. This is a large Black-red male.

▲ The eyes are pearl-coloured and set at the side of the head, giving the breed a sinister appearance. It has a characteristic, powerful beak.

holds its tail at an angle sloping to the ground, and the tail feathers fan outwards. It has long legs, thin thighs and powerful leg muscles. The Aseel has a strong beak, small earlobes and a pea comb; wattles are absent. A unique breeding regime in its country of origin bestowed courage and stamina upon the breed with an inbuilt inclination to fight. Both male and female birds will fight to the death. Even young chicks will attack each other at just a few weeks old, and may also attack their mother. As such, they should only be put with other non-fighting breeds; and males should be separated from each other. Despite the inclination of the young to fight, Aseels make good mothers, but are bad egg-layers. Along with other strains of breeds developed in isolation from mainline poultry, the Aseel brought hybrid vigour to the western poultry world. It has high muscle and meat content.

Rare and even obscure, many named strains of Aseel can be found at specialist shows, and the purest is likely to remain with specialists.

▶ The Aseel has characteristics rarely found in other poultry breeds. For example, the duck foot shown here is allowed, but is considered a serious fault in any other breed.

ESSENTIAL CHARACTERISTICS

Size: Male to 2.4kg/5½lbs. Female to 1.8kg/4lbs.
Varieties: Black-red, Black, White.
Temperament: A fighting bird, strong, muscular and hardy. Friendly towards humans.
Environment: Tolerates confinement or free-range conditions.
Egg yield: 20 tinted to brown eggs per year. They are large for the size of bird.

Shamo

The Shamo is indigenous to Thailand and was developed in Japan, where a number of forms exist. It is a protected breed in Japan and is that country's best-known game bird. Cockfighting is still legal in Japan, and the bird is bred there for fighting purposes. The name Shamo is a derivation of the Japanese word for Siam, the original name of Thailand. The Shamo breed name is used to describe several breeds, ranging from O Shamo, which is the tallest, through to Chu Shamo, Ko Shamo and Nanking Shamo. The latter two are bantams. Over the years, the Japanese have developed several other obviously related breeds, including the shorter and grotesque Yamato Gunkei. While the latter two are treated as bantams in Britain and Europe, in their homeland Japan they are seen not as

► *Like the Aseel, the Shamo has a pea comb, small lobes and barely has wattles. This is the Ko Shamo.*

bantams but just as a smaller breed in a range of breeds that are each bred to conform to its own standard and weight range. This will help explain the wide range of sizes and weights often found among Ko Shamo entered in bantam classes in the West.

The breed reached Europe from the USA 40 years ago, where all of this group, as they appear in their present form, have been kept solely as exhibition fowl. The Shamo breed has been used to improve the table qualities of other breeds.

The Shamo is the tallest breed of poultry, with males regularly reaching 76cm/30in tall. The breed has an upright habit, with a long sloping back, broad breast, square shoulders and attenuated neckline. While it is similar to the standard-bred Malay, the Shamo is without the pronounced curves of that breed. It also has a thinner pea

ESSENTIAL CHARACTERISTICS

Size: O Shamo Male 3.6–5.4kg/ 8–12lbs. Female 2.7–4.5kg/6–10lbs. Chu Shamo Male 3kg/6lbs 10oz. Female 2.2kg/4lbs 14oz (minimum). Nanking Shamo Male 1.1kg/2lbs 8oz. Female 737g/26oz. Ko Shamo Male 1.1kg/2lbs. Female 793g/28oz.

Varieties: Black, White, Black-breasted Red, Spangled, Dark, Brown-red, Wheaten.

Temperament: Docile, but males can be belligerent.

Environment: Tolerant of some confinement as adults but adolescent fowl need space to exercise.

Egg yield: Poor, 20–40 per year, but bantam varieties could lay more.

◄ *A Black-red Shamo male. While superficially similar to the breed standardized as Malay in the Western world, the Shamo has a pea comb and flat back; features that are closer to those of the Kampong fowl of Northern Malaysia.*

comb. This breed has particularly short and hard feathering, with the plumage failing to cover the breastbone and sometimes the shoulder area, and the face has harsh skin. Shamos look somewhat sinister with a medium-length hooked beak and pale orange eyes. The skin is red.

Breeders wishing to rear the Shamo for the sheer pleasure of keeping and conserving examples of some of the most unusual and rare breeds have to be constantly on their guard against those who try to steal them in the mistaken idea that they could still be used for cockfighting.

Malay

The Malay is a very striking bird, with long legs, a powerful stance, broad chest, long neck, gaunt appearance and cruel expression. These are birds to be wary of and are unsuitable for keeping as a family pet. These are instinctively aggressive fighting birds. Two roosters in a pen may fight to the death and inflict significant damage. A rooster may also become aggressive to hens. Because of their size Malays need room to roam, and they regularly dig holes. In its native region, the Malay is regarded as an ancient village bird that roams freely.

By about 1870, examples of these tall Asian Malay fowl had found their way to Britain and in particular to Cornwall, the first stopping-off point for ships returning from Asia. Malays made little impact until they were seen at the early poultry shows. Although standardized as

ESSENTIAL CHARACTERISTICS

Size: Large male 4.9kg/11lbs.
Large female 4kg/9lbs.
Bantam male 1.1–1.3kg/40–48oz.
Bantam female 1–1.1kg/36–40oz.
Varieties: Black-breasted red Male, Clay, Cinnamon, and Red Wheaten Female. Creel, White, Black.
Temperament: Aggressive, but can be docile when regularly handled.
Environment: Free range. Requires room to exercise.
Egg yield: 30 medium light brown eggs per year.

Malay, the standard requirements of the neck, back and tail forming three curves, plus the walnut comb, suggests that most of these early imports were the taller strains native to states of British India rather than those states that are now part of Malaysia. As an exhibition fowl, the Malay was bred to an agreed standard by 1880, one which we would still recognize today. Its appearance is likely to owe much of its development to British, and, more specifically, to Cornish fanciers. It seems highly likely that both large and bantam Malays in their standardized form are a product of Cornish breeding.

The huge reach and weight of the large Malays may help to explain why some of the miniature versions are the largest of their class seen on the show circuit. As a breed type they may have donated many useful traits to the later heavy and specialized table breeds

▼ A *Crele male: note the curved neck and back. The tail provides the third similar curve.*

and strains. However, as exhibition-bred birds, both large and bantam versions are generally appalling layers. The older the hen, the fewer eggs she is likely to lay. Malays are also noted as bad mothers, in part because of their aggressive nature. They may inadvertently kill their young in the process of attacking another female.

The heavy weight and height of the full-size Malay will inevitably mean that its miniature version can be expected to be one of the largest of the bantams. As in the large version, reach is considered to be one of the most important breed points. Exhibitors tend to exhibit and breed from their tallest examples, which has led to over-sized examples being found both in Malay bantam show classes and breeding pens.

▶ *A young Brown-red Malay male. Its tail has yet to develop the characteristic third curve.*

Modern Game

Within a few years of the various Acts of Parliament banning cockfighting, Britain became gripped by a craze for exhibition poultry. Breeders looked to exhibit their former cockfighting birds rather than use them for sport. Old English Game birds, which were the pit breed, were not judged appropriate for showing. Exhibition judges looked for birds that were taller than Old English Game, with a shorter back, more upright carriage and hard feathering. Initially, to increase the size of their game birds, breeders crossed them with the Malay poultry breed. With greater height, "reachiness" became a feature of the birds used to attract points at exhibitions. Fanciers managed to breed out the clumsiness of the Malay breed from these taller game birds, and produced a very elegant fowl. The scanty feathers and tail carriage of the new breed undoubtedly came from the Malay, however. It was with the formation of the Old English Game Club that these newly bred fowl become known as Exhibition Modern Game to distinguish the breeds, though later the word "Exhibition" was dropped from their name.

The popularity of Modern Game peaked by the early 1900s, after which Old English Game enjoyed a resurgence of popularity at the expense of Modern Game. The large Modern Game continued to lose popularity until its nadir in the 1970s when it could have been considered extinct. For the last 40 years, a small group of enthusiasts have largely replicated the efforts of the previous century, reinstating the bird to its full exhibition status. Very few examples are seen at today's shows, and because of this they are a fascinating reminder of the path that exhibition selection can take a breed. The birds are always likely to remain scarce, but will appeal to those who enjoy the competitive aspect of game exhibition.

The game bantams that were developed probably included smaller strains of the earlier cockfighting game, as well as some of the smallest crossbred bantams. Modern Game bantams remain one of the most important, if challenging, bantam breeds. Victorian breeders would have wanted both large and bantam versions to be seen as exhibition birds, and while only standardized in seven colours at that time, they would have required each of these to conform to the most exacting rules for plumage pattern, eye and leg coloration. This range of colours has been extended by modern breeders. The bantam version remains in the hands of specialist breeders.

All Modern Game birds have yellow or green legs. Males are dubbed on reaching adulthood. This is a Brown-red rooster.

▼ *This Black-red Modern Game rooster is an excellent example of breed type. It is tall and slimline with a neat, compact body and short flat back. The bird has an upright, lofty carriage, and stands on long, tall legs. Its tail is whip-like.*

ESSENTIAL CHARACTERISTICS

Size: Large male 3.1–4kg/7–9lbs.
Large female 2.2–3.1kg/5–7lbs.
Bantam male 566–623g/20–22oz.
Bantam female 481–510g/17–18oz.
Varieties: Black, Blue, Birchen, Black-red, Blue-red, Brown-red, Gold-duckwing, Lemon-blue, Pile, Silver-blue, Silver-duckwing, Wheaten, White.
Temperament: Quiet, but exhibition birds will require careful handling.
Environment: The tiny bantams will stand confinement but may be less robust than other game bantams.
Egg yield: Up to 100 per year. Bantam eggs are tiny.

Old English Game Bantam

By far the most popular British exhibition bantam, Old English Game bantams would have been practically unknown until the late 19th century, and despite their name, would have evolved from the existing tall exhibition bantams that were later to become known as Modern Game. By the 1920s, photographs depict them as miniatures of the Oxford show game. In the UK a specialist national show is devoted to this breed. However, unlike most other breeds, a long period of separate development has seen these game bantams being treated as a completely separate breed.

As long as standard Old English Game birds remained on farms, country estates or smallholdings with enough room to display their natural athleticism, most flocks would have remained very like their pit-game ancestors. By nature, they are bold, proud and agile. The bantam version may initially have lacked some of the muscle and feather-hardness of the standard size. However, because of

▼ A Furness or Black-red male.

their diminutive size, they were far more likely to attract breeders with limited space for the birds to range. These bantams soon adapted to being kept in tiny pens that allowed their owners an easy opportunity to handle these small birds and assess firmness and shape. Before long, they had been bred to have the hard-feathered quality of their large counterparts. Other characteristics of the large Old English Game breed were also incorporated into the small breed. Some strains soon developed to a point where

▶ A Spangled male showing the sort of bend to its hock that is sometimes absent even in exhibition examples.

◀ When breeders started to select bantam versions of Old English Game, the birds would have looked more like this Ginger male. Such types are now occasionally exhibited as miniature Oxfords.

ESSENTIAL CHARACTERISTICS

Size: Male 680g/24oz.
Female 623g/22oz.

Varieties: Barred, Birchen, Black,
Black-breasted Red, Black-tailed Red,
Black-tailed White, Blue, Blue Brassy-
back, Blue Golden-duckwing, Blue
Millefleur, Blue Quail, Blue-red, Blue-
wheaten, Brassyback, Brown-red, Buff,
Columbian, Crele, Cuckoo, Fawn,
Fawn-breasted Red, Fawn silver-
duckwing, Ginger-red, Gold-
duckwing, Lemon-blue, Mealy-grey,
Millefleur, Mottled, Porcelain,
Quail, Red, Red-pile, Red Quill,
Self-blue, Silver-blue, Silver-duckwing,
Silver-quill, Spangled, Splash,
Wheaten, White.

Temperament: Jaunty. The more
developed strains are not good
at flying.

Environment: Will adapt to free
range. The exhibition strains that have
been kept for generations in small
runs seem happier than most fowl.

Egg yield: 80–100 very small tinted
eggs per year.

they differed from the Oxford type,
with those handling them beginning
to describe their body shape as that of
a "flat iron or bullock heart".
Differences developed and the hard
feathering led to a tighter and smaller
tail. The breed standard asked for
"wings held low to protect the
thighs". However, a concentration on
overall body shape soon saw a higher
wing carriage become the norm for
the bantam type. This in turn led to
many breeders regarding "handling"
as the most important aspect of the
breed. Breeders who kept their birds
confined to small spaces often owned
birds that were more soft-feathered
and fluffier than their large
counterparts. The birds were
continually assessed for the hardness

of their feathers and overall body
shape. As more breeders favoured and
bred from birds that had a "flat, iron-
shaped back and bullock heart-
shaped body", the
birds ceased to
look and behave like
miniature versions of
Oxford Game. Later, as large
Carlisle Game became popular, many
saw Old English Game bantams as
being their miniatures.

During World War II, large game
birds almost disappeared from farms
and smallholdings. Post-war food
shortages meant that most poultry
was kept solely for egg or meat
production. Few could find room and
precious food to keep a few tiny Old
English Game Bantams.
At exhibitions there
have always been
classes for Old
English Game
bantams. For most
exhibitors, their
interpretation of type, shape
and feathering has been far more
important than the colour of the
bird. Some judges agreed, and often
birds of no fixed or "off" colour were
promoted to best in show.

▼ *A Furness hen.*

◄ *A Silver-duckwing male showing the
same sort of front and short back seen
in some large Carlisle Game on today's
show circuit. Note the wing carriage of this
bird when compared to the Oxford type,
particularly the Ginger male opposite.*

Indian/Cornish Game

Also known as the Cornish poultry breed, these fowl are the most solid and muscular looking of all poultry, with wide-set, strong, short legs, a deep-set breast and a solid appearance. The breed was developed for fighting and contains Old English Game, Malay and Aseel/Asil in its genes. However, it is a slow-moving bird, and its thick-set appearance means it was never agile enough to fight competitively, and although unsuitable for fighting, the bird was developed for the table instead. The massive size of the bird makes it particularly suitable for eating, and often it is crossed with the Sussex or Dorking breed for this purpose. Indian Game are poor egg-layers. Most cannot breed naturally because of their enormous body size and short legs, consequently they do not make good mothers.

Indian Game have yellow legs and flesh, which was once considered undesirable for the top end of the poultry meat market. They are slow-growing and only modest egg-layers, and are therefore undesirable as a commercial prospect.

ESSENTIAL CHARACTERISTICS

Size: Male 3.2–4.2kg/7–9lbs. Female 2.7–3.6kg/6–8lbs.
Varieties: Cornish, Jubilee.
Temperament: Vigorous, active.
Environment: Back garden.
Egg yield: Up to 100 tinted eggs per year.

However, like their Aseel ancestor, when crossbred, Indian Game are capable of passing on useful genetic traits, such as abundance of breast meat. This in turn enabled poultry geneticists to incorporate Indian Game genes into some of the most successful meat birds found on our supermarket shelves.

In spite of their ancestry, the Indian is a docile and friendly

▲ *A large Jubilee rooster.*

fowl and makes the perfect pet. The bantam versions are exact miniatures and are suitable for a small garden.

THE CHANTECLER BREED
Indian Game was one of five breeds used to create the Chantecler breed, a dual-purpose fowl able to withstand the hard winter frosts of Canada. The initial cross was of a Dark Indian Game male with a White Leghorn hen, and then a Rhode Island Red male with a White Wyandotte hen. Each cross produced white or off-white pullets with tiny pea combs that were thought desirable in extremely cold conditions. The offspring of these two crosses were mated. Some of the very best resulting pullets were later mated to a White Plymouth Rock. In total it took nine years to produce the perfect Chantecler breed. While for a time widely kept by Canadian farmers, the breed is now rare.

▶ *A Dark Indian Game bantam hen.*

Rumpless Game

Few breeds of poultry have a rumpless variety, and most are considered to be quite rare. Rumplessness, where the breed effectively has no tail feathers and instead has a rounded rear, is caused by a genetic defect in which the end of the vertebrae and what is usually known as the parson's nose, from which tail feathers grow, is missing. The oil glands that would serve the tail feathers are also missing.

Rumpless game are a tailless version of the Old English Game bird. They are available as large or bantam versions, though the large versions are much rarer. They are also known as Manx Rumpies, and are thought to have descended from

ESSENTIAL CHARACTERISTICS

Size: Male 2.2–2.7kg/5–6lbs.
Female 1.8–2.2kg/4–5lbs.
Bantam male 623–737g/22–26oz.
Bantam female 510–623g/18–22oz.
Varieties: All game colours. Few birds have identical coloration.
Temperament: Game characteristics.
Environment: Free-range.
Egg yield: Ornamental value.

Persian rumpless game, although the exact origins appear to be lost.

Rumpless game birds have an upright posture, a forward-thrusting carriage and a rounded body that slopes down and back. A wide variety of colours are known, and it is unusual to have two identically coloured birds. The breed has a single comb, red earlobes and no standard leg colour. While the bantam version is virtually a rumpless version of the Old English Game breed, they are not customarily dubbed, and should, when exhibited, be entered in the rare breed classes. They may, however, be found to be less quarrelsome than many of their standard feathered counterparts.

◀ *This is a Blue-red male bird.*

▼ *A Rumpless Game. Also known as Manx Rumpies and Persian Rumpies, they are thought to originate from what is now Iraq.*

▶ *A Blue-wheaten bantam hen.*

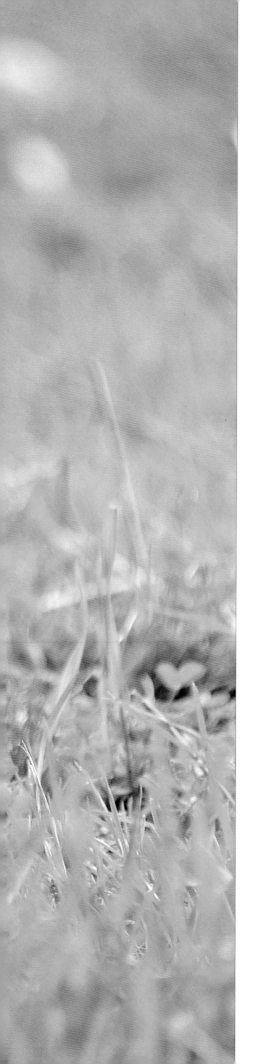

A DIRECTORY OF TRUE BANTAMS

The defining characteristic of a true bantam is that, unlike other poultry breeds, it does not have a large counterpart. Smaller versions of large breeds are often incorrectly known as bantams, but should in fact be termed miniatures. The true bantam classification covers a handful of breeds, each with distinguishing characteristics, which are admired by many breeders. These small birds are often available in a wider variety of stunning colours, and their visual appearance can be immensely appealing, adding ornamental value to the garden.

Such small birds make good pets for young children. In addition, true bantams are the perfect pets for those with limited space; they require smaller housing, taking up less room. They also inflict less wear and tear on the garden environment. These birds have lively and perky characters, with a gentle and friendly nature. They are feisty and can be extremely entertaining to watch. True bantams lay eggs that are smaller than those of standard hens, providing a practical purpose for keeping them as well. They will need to be well protected from predators such as cats and foxes.

▲ *True bantams have a bold but gentle disposition and respond well to humans.*

◄ *Available in a wide range of colours, the plumage of true bantams is often appealing to breeders with an interest in perfecting unusual varieties.*

WHAT IS A TRUE BANTAM?

Weighing no more than 750g/1½lb, these small birds are the dwarves of the poultry world. For all their small size, these are fowl with large personalities, lively dispositions and friendly natures, making them suitable pets for those with less room to spare.

In 1970, the term "true bantam" was adopted to describe all poultry breeds that do not have a large version. All of the true bantams, standardized as such throughout much of the world, are thought to have descended from an ancient population of tiny fowl that may have existed for centuries in the Javanese islands of Indonesia, before being spread through both East and West along the trade routes. The term "bantam" derives from Batavia, the Dutch name for Jakarta, a port on the island of Java, which was used by the Dutch East Indies Company to trade with Europe. It is likely that sailors brought the small fowl that they found in this location back to Europe; these original bantams would then have naturalized in Northern Europe. It is plausible that imported true bantams were crossbred with existing undersize

strains of English game or Hamburg-type breeds. Elegant strains of miniature game fowl, far removed from the early pit game, would seem likely to have developed almost entirely from these bantams.

Nowadays, the word bantam has become the accepted term for any tiny or diminutive fowl. Small

◀ Belgian bantams are bright and perky by nature.

versions of large fowl are often wrongly known as bantams, but these are in fact miniatures, despite the fact that many miniature versions of large breeds can be anything but miniature in size. True bantams were used in some initial crosses that formed the foundation of many miniature versions of large fowl breeds.

Bantams are ideal for poultry keepers with limited space. Their food requirements are similar to those of other poultry but in smaller quantities. Their eggs are half or even a third of the size of those of standard breeds.

Group characteristics

All members of the true bantam group have a different shape to those of miniature versions of large fowl. They characteristically have very short backs and a low wing carriage. In some instances the wing carriage is almost perpendicular to the back of the bird and the wings nearly touch the ground.

Group classification

The Victorian craze for exhibition poultry led to the concept of breeding fowl to conform to a written standard. As a result, breeds were grouped according to type. Apart from game breeds, the early administrators chose the terms "sitters" and "non-sitting breeds", referring to the breeds' inclination, or not, to sit on their eggs

◀ True Belgian bantams have many characteristics of standard breeds including muffs and beards.

▲ *Belgian bantams come in a range of colours practically unique to the group of breeds.*

▲ *With their rounded body shapes and attractive personalities, Pekins make appealing pets.*

▲ *Feather-legged Booted bantams need to be kept in an environment where the foot feathers will not be damaged.*

and rear young, as the classifications of choice. Shows classified miniature versions of sitters and non-sitters as variety bantams. Today, breeds are classified as heavy and light, according to weight, rather than as sitters or non-sitters, with plenty of categories within each class to further distinguish birds by type and by origin.

For most of the 20th century, the breeds that had partly or wholly derived from the Javanese true bantams were listed along with Pekins/Cochin bantams, Poland bantams and so on, as "ornamental bantams". By the 1970s, the miniature versions of the soft-feather breeds were classified as dwarves, and

there was a growing awareness of the need for a system that classified all the bantams derived wholly or largely from the original Javanese imports as true bantams.

The breeds

Today, the breeds that make up the true bantam classification have been selected and perfected in different areas of the globe. Like some large breeds, there are true bantam breeds with feathered feet, beards and muffs, long saddle, hackle and sickle feathers, distinctive feather patterns, modifications and coloration. There are a variety of combs, and different stances. There is even a breed with short legs, and a few that are rumpless. In fact, all the true bantam breeds are easy to distinguish from each other because their appearances are so individual. Many make good broodies and will look after the eggs of other poultry breeds.

◄ *Belgian bantams make good pets for people who want eggs but have little space. They are available in a huge range of colour varieties.*

Sabelpoot/Booted Bantam

Those Booted Bantams shown at exhibitions today are likely to be descended from Continental strains, principally from the Netherlands, where they have long been standardized as Sablepoot. Yet fowl similar to Booted Bantams were referred to in British books around 1850, exhibiting a fuller tail than those currently being shown as Sabelpoots. The Booted Bantam hails from the Netherlands where it is known as the Sablelpoot or Sapelpoot. This is an old breed. Booted Bantams are historically and genetically linked to the feather-footed Belgian bantams.

▼ *A White male.*

They were was crossed with the Barbu d'Anvers breed to make the Barbu d'Uccle breed. In their homeland the breed is relatively common. True to its name, the main distinguishing feature of this breed is the extravagant feathering on its feet. The breed has a majestic appearance, with an upright stance, deep curving chest, wide and upright tail feathering and a single red comb, full wattles and red earlobes. It is available in a wide number of varieties, all of which are attractive. As a result of the feathering on the feet, this breed is less likely to cause damage to flowerbeds than other breeds when left to range free in a garden environment. Booted Bantams do have a foraging nature, however. These birds have a calm and friendly temperament and make great family pets.

Booted Bantam genes were used to create Cochin bantams in 1884. A descendant of their progeny was crossed with American Cochin bantams in 1890. It is a measure of the way true bantams have helped to create miniature versions of the large breeds.

▼ *A Buff-mottled hen.*

◄ *The feet feathers can easily become damaged if not looked after. A deep litter hen house may help to protect the feathering.*

ESSENTIAL CHARACTERISTICS
Size: Male 850g/30oz.
Female 750g/27oz.
Varieties: Black, Blue, Cuckoo, Millefleur, Mottled, Porcelain, White.
Temperament: Docile.
Environment: Needs dry housing to protect the feathered feet from wet weather.
Egg yield: 80–100 eggs per year.

BELGIAN BANTAMS

The Belgian bantams, made up of the Barbu d'Uccle, Barbu d'Anvers, Barbu d'Everberg, Barbu de Grubbe and Barbu de Watermael are the only true bantams to be standardized with two or more ornamental appendages. These small fowl with feathered legs are thought to have found their way to northerly regions of Europe from Asia during 300 years of trade with the Dutch East Indies Company. The original imports are thought to have been beardless and muff-less, and it seems likely that today's breeds are derived from the careful selection and crossing of their descendants with other tiny bantams brought back

to the same region. The feather-legged Barbu d'Uccle and clean-legged Barbu d'Anvers quickly found favour, while the rumpless Barbu de Grubbe and Barbu d'Everberg were largely ignored. Along with the crested Barbu de Watermael, these three breeds remained minority varieties for many years, even within their homeland. Four of the five breeds are standardized with the same colours. Exhibitors need to understand all the details of their most complex colour patterns, as well as the shape of every variety. Belgian bantams make the best of pets and are perfect "chatty" garden companions.

Barbu d'Uccle

The Barbu d'Uccle, also known as a Booted Bantam, is characterized by its feathered feet and beard. The hen has a muff and beard, with the features forming distinct tufts. Both male and female carry a neat, and relatively small, single red comb. This breed has abundant feathering, with the hackle feathers reaching down to the saddle area of the back. The males have a tail sward that is carried in an open configuration. The breed has a low posture with wings that are also carried low to cover the vulture hocks. All these features combine to create a breed with a proud stance.

The breed has a gentle and docile

nature. It is placid and undemanding, except during the breeding season, when males have been known to attack humans, other birds and animals. The hens go broody and make good mothers.

▼ *There are strong historical connections between colour and variety. Many people wrongly assume that the Millefleur colour is specific to this feather-legged breed. This is a Millefleur rooster.*

ESSENTIAL CHARACTERISTICS
Size: Male 800g/28oz.
Female 550g/19oz.
Varieties: Black, Blue, Cuckoo, Lavender, Millefleur, Porcelain, Quail, White.
Temperament: Tolerant.
Environment: Care is needed to ensure that the leg feathers do not get broken.
Egg yield:
80–100 white
or tinted
eggs per
year.

▼ *Male birds are tolerant toward each other, and can be housed together except in the breeding season. This is a black rooster.*

Barbu d'Anvers

The stance of the Barbu d'Anvers is distinctive. It is that of a bird standing to attention and showing off, with the chest puffed out and upright. The male carries his tail at a perpendiuclar angle and the wing and wing feathers point downward almost covering the clean legs. Belgian d'Anvers are capable of flying. The breed has thick hackle feathers, which in the male look like a mane and in the female like a ruffle. The body is short. The males have a tail sward rather than sickle-shaped feathers. Like all the Belgian bantams, this breed has a crest and beard that suppresses the wattles. It has a broad rose comb that ends in a leader. This is an attractive breed that makes a good pet, and fowl will become tame with handling.

▲ *A Quail hen.*

▶ *Some of the colour varieties of Belgian bantams can be difficult to perfect.*

ESSENTIAL CHARACTERISTICS
Size: Male 680–790g/24–27oz.
Female 570–680g/20–24oz.
Varieties: All of the standard Bearded Belgian colours.
Temperament: Cheerful, chatty.
Environment: Can be kept in a small pen.
Egg yield: 80–100 white or tinted eggs per year.

Barbu d'Everberg

The Barbu d'Everberg is regarded as a rumpless subvariety of the Barbu d'Uccle breed. Like all rumpless varieties, it would have originated from spontaneous sports of fowl with standard tails, in this case a Barbu d'Uccle. Rumpless breeds are few and far between in the poultry world. This particular breed was first recorded in 1904. The rumpless gene is a dominant one, so whenever the Everberg is crossed a rumpless form will always result. Where the rumpless characteristic is found to be the result of a dominant sport or mutation, it offers the possibility of creating new colours. Rumpless breeds have the last two vertebrae missing. It is to these vertebrae that the tail feathering is usually attached. The rest of the body is covered in abundant feathering. The back is covered in long saddle feathers; the feet also have feathers. This breed has a crest and beard, small single comb and no wattles. The rumpless feature is appealing.

◀ *The characteristic rounded boule is as much a part of the Barbu d'Everberg as its beard and muff.*

ESSENTIAL CHARACTERISTICS
Size: Male up to 800g/28oz.
Female up to 600g/21oz.
Varieties: All the standard Bearded Belgian colours.
Temperament: Perky.
Environment: Can be kept in small pens. Requires a dry floor area.
Egg yield: 80–100 white or tinted eggs per year.

Barbu de Grubbe

This is another rumpless true bantam, of the Barbu d'Anvers genetic line. This breed has a soft and gentle temperament, though not in the breeding season, when males are known to become aggressive.

The Barbu de Grubbe is a small breed that has a pert, upright stance and wings that slope to the ground. It has a short, thick neck, a beard and muffs, single red rose comb, and a wide, rounded rear body shape. As in Belgian bantam varieties with beards and rose combs, the de Grubbe should not have any wattles. A similar but probably far larger fowl illustrated by Aldrovandus (an Italian naturalist regarded as the father of natural history) in 1600 was labelled as a Persian fowl, and later, rumpless fowl were recorded near Liège in Belgium, identified as hedge fowl. These were valued because, having no tails, they were more likely to escape from predators such as foxes.

ESSENTIAL CHARACTERISTICS
Size: Male 650g/23oz. Female 550g/19oz.
Varieties: All Belgian bantam colours.
Temperament: Perky Belgian character.
Environment: Can be kept in small runs and houses.
Egg yield: 80–100 white or tinted eggs per year.

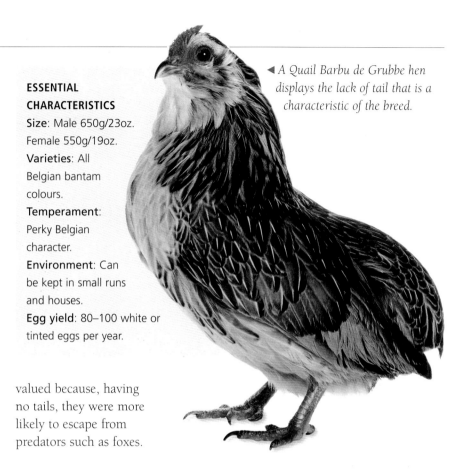

◄ *A Quail Barbu de Grubbe hen displays the lack of tail that is a characteristic of the breed.*

Barbu de Watermael

ESSENTIAL CHARACTERISTICS
Size: Male 600–700g/21–24½oz. Female 450–550g/16–19oz.
Varieties: All Belgian colours.
Temperament: Perky, cheerful.
Environment: Small runs and houses.
Egg yield: 80–100 white or tinted eggs per year.

The Barbu de Watermael is the smallest Belgian bantam. It has a small square rose comb, which is unique in having three leaders. It has a tri-lobed beard and clean legs, and is the only member of the bearded Belgian bantam family with a crest. The crest may in some way inhibit the development of the characteristic boule. The unique tri-lobed beard adds its own distinctive feature.

▶ *This breed has clean, unfeathered legs.*

◄ *A Lavender Quail male with a unique tri-leader comb.*

Dutch Bantam

The tiny Dutch Bantam is one of the smallest of all the true bantams, and is the most popular poultry breed in its native Holland. It is probably descended from a Dutch native breed, but may include Indonesian ancestry. The Dutch East Indies trading company had closer and longer trading links with Indonesia, and it is likely that sailors included poultry on their vessels to provide food as well as for trading. Asian poultry brought back to Europe by these sailors could have become absorbed into local landrace bloodlines. Because local landrace poultry was commonplace, the Dutch Bantam was not recognized offically as a breed until 1906.

This is a breed that lays tiny eggs, and it may have been selectively bred for this characteristic. It is believed that in past centuries, large eggs had to be passed to the kitchen of the local landed gentry, and farmers were only allowed to keep those that were small in size.

▲ *A Silver-blue Partridge hen.*

Dutch Bantams found their way to Britain in 1970, but appeared much earlier in the USA. The breed died out in the USA due to lack of interest, but has since been successfully reintroduced. Breeders soon established a club and strong show presence. Dutch Bantams often win top awards when shown competitively.

This tiny fowl is a perfect miniature, free from any of the appendages and exaggerations that are essential to the anatomy of many other true bantams and ornamental fowl. Unlike the Rosecomb breed, the earlobe of the Dutch

ESSENTIAL CHARACTERISTICS
Size: Male 550g/19oz.
Female: 450g/16oz.
Varieties: Birchen, Black, Blue, Blue-partridge, Blue-yellow partridge, Buff Columbian, Crele, Columbian, Cuckoo, Gold-partridge, Lavender, Millefleur, Mottled, Partridge, Pile, Quail, Salmon, Silver-partridge, Silver-blue Partridge, Yellow-partridge, Wheaten, White.
Temperament: Flighty, friendly.
Environment: Back garden.
Egg yield: 160 cream or light brown small eggs per year.

Bantam is small and unexaggerated. It has slate-blue legs and relatively large wings for its size. This is a breed that can fly. The modest single red comb is not expected to sit tight to the head as in most true bantam breeds, but is allowed to "fly away" slightly. Initially bred in the Black-red and Silver-partridge colours, the varieties now number at least 15. Their colours are thought to be the exact counterparts of rudimentary game bird patterns. All are carefully standardized colour combinations that make the breed one of the most colourful additions to the show bench or garden.

By nature the Dutch Bantam is alert and flighty. The breed has a friendly disposition, making it a good bird to

▼ *A Black male.*

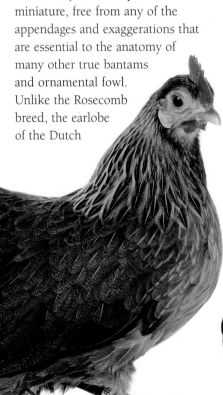

► *A Gold-partridge hen.*

▲ *A Lemon-porcelain male.*

◄ *A Gold male.*

▼ *A Gold-partridge male.*

The Dutch Bantam adapts to being kept in a relatively confined space, but like all bantams kept in close confinement, regular interaction with humans is essential for the wellbeing of the bird. The breed is hardy and remains one of the least complicated and most satisfactory tiny garden fowl.

keep as a pet. In addition, it lays a good number of eggs considering its small size, which makes it a practical breed to keep. The breed is broody, and is known to make a good parent.

Nankin

Nankins are thought to be the result of early crosses between Javanese imported poultry and some of the small strains later standardized as Hamburgs. However, there is also speculation that the breed hails from Nanjing in China. Nankin or nankeen was also an early description of the colour of a fabric imported to the West from the Far East.

This is an old breed thought to date back to the 1700s, and is considered by some authorities to be one of the foundation breeds used to establish Sebright bantams.

Nankins are only found in a natural two-tone form of buff colour, and are available in two varieties, single comb and rose comb. They made little impact

▲ *A single comb male.*

▲ *A Buff rose comb hen with a natural two-tone pattern.*

ESSENTIAL CHARACTERISTICS
Size: Male up to 680g/24oz.
Females up to 625g/22oz.
Varieties: Single comb, Rose comb.
Temperament: Cocky, bouncy, less placid than other true bantams.
Environment: Will be happy to range free.
Egg yield: 120–140 smallish tinted eggs per year.

on the exhibition scene and had more or less disappeared until they were rediscovered in Britain in 1970. The breed remains under the protection of the Rare Poultry Society in Britain. The Nankin retains the type, bounce and character of all true bantams and is probably as good an example as we are likely to find of yesteryear's bantams of the countryside.

Japanese/Shabo

This is an ancient breed, recorded more than 1000 years ago in Japan. However, for centuries Japan was not open to trade with the western world, so the breed did not arrive in Europe or America until the 19th century.

This is a distinct bantam breed. The breed standard requires a precise tail carriage that in the males is somewhere between upright and veering beyond perpendicular to the body. It also has extremely short legs and weak thighs. An interest in the grotesque may have led Japanese breeders to select for and accentuate the shortening of leg bones. In Europe, dwarfism had been noted in Scots Dumpy populations. Like this

ESSENTIAL CHARACTERISTICS
Size: Male 510–600g/18–20oz. Female 400–510g/14–18oz.
Varieties: Birchen-grey, Black, Black-tailed Buff, Black-tailed White, Blue, Brown-red, Cuckoo, Mottled, White.
Temperament: Placid.
Environment: Dry, protected.
Egg yield: 60–120 small cream or white eggs per year.

▶ *A Grey male. Note how the tail is held upright even when the bird is sitting.*

European breed, the Japanese bantam also suffers from a lethal creeper gene. Both parents have the long and short leg gene. If both pass the short leg gene to the offspring, those embryos will fail to thrive. This happens in 25 per cent of the offspring. Another 25 per cent will be born with long legs. This feature alone makes this fowl a difficult breeding proposition.

The breed has a deep and full breast, a short round body, long

▼ *A White frizzled hen.*

saddle feathers, and wings that touch the ground. It also has some unique and challenging colour combinations. Because of its short legs the breed has a low carriage, and requires protection from cold and wet.

The last few years has brought greater movement of breeding lines out of Japan, along with a better understanding of the culture behind the breed. This same movement of ideas and stock has seen more frizzle-feathered examples, as well as the emergence of a silky-feathered variety. The Japanese bantam is one of the most important exhibition breeds. However, the breed is always likely to remain in the hands of experienced breeders and dedicated exhibitors. Like so many of the true bantams, where breed type and style have developed in the hands of fanciers and exhibitors, the Japanese bantams are content to spend much of their life in reasonable confinement. The breed is reliant on its owners and is happiest when given plenty of attention.

Rosecomb

Rosecombs are an old breed, which may have been developed in England in the 15th century. They are kept primarily for exhibition or ornamental value in the garden, since they are known to be poor egg layers and not worth keeping for their meat. Exhibitors probably spend more time preparing their birds for show than breeders of any other fowl.

Early illustrations depict black and white varieties with single or rose combs and either red or white lobes. However, the term "Rosecomb" was once used to describe several breeds that had a rose comb, rather than one specific breed. English breeders were largely responsible for refining the breed, introducing Hamburg genes to improve the quality of the feathering. When shown beside a Hamburg, the two breeds do have similarities. Perfect rose combs and smooth white

▼ *The white lobes are smooth with a velvety texture. This is a black male.*

lobes were developed to become the most important defining features of the breed. By the turn of the 20th century, Rosecombs had reached a stage of development that would be recognized by today's breeders.

Black, White and Blue are the most common varieties since breeders concentrated

on perfection of head points rather than colours. In profile, the breed has a short back, arched neck and proud, upright stance. The tail has wide feathers and is carried upwards at an angle, slightly higher in the male than the female. The wings are held low and at an angle but do not obstruct the leg visibility.

British Rosecombs are almost identical to those shown in Holland as Javas.

ESSENTIAL CHARACTERISTICS
Size: Male to 600g/21oz.
Female to 550g/19oz.
Varieties: Black, Blue, White – other colours are recognized, though rarely seen.
Temperament: Hardy, active, good flier, easily handled.
Environment: Tolerates confinement.
Egg yield: Small, cream-coloured. Egg numbers can be poor; 50–80 per year.

▼ *The comb may be disproportionately large in the Rosecomb breed. It should have a square front with the central area covered in protrusions. The long tapering leader leans back.*

◄ *Rosecomb hens rarely go broody and may be difficult to breed. Chicks may have a high mortality rate either before or just after hatching. In addition, males may have low fertility rates. This is a Black hen.*

Pekin/Cochin Bantams

In Britain the breed of bantam called Pekin and treated as true bantams, is known as the miniature Cochin throughout most of the rest of the world.

A single pair of Pekin bantams was sent to Britain with other booty after the sacking of Peking's (Beijing's) Summer Palace by Anglo-French forces during the Second Opium Wars of 1860. At that time, any poultry from distant regions was thought to be the indigenous breed of that district, hence the naming of the breed for its point of origin. By 1863 the offspring of this single importation were out-crossed with a White Booted Bantam. With just this one line to breed from, some breeders resorted to out-crossing the breed with Nankin bantams. All the breeds used to create the early Pekins would have been true bantams. W. F. Entwistle, who is generally credited with making many early miniature versions of large fowl, used these birds to cross with others that he had imported from Shanghai in the 1880s. By 1890 he co-operated with American breeders, loaning one Buff rooster which was reported to have sired 30 chicks. Reading his notes, there can be little doubt that the intention was to create miniature Cochins. All the illustrations that he made show birds that are exact

▼ A Blue male with long saddle and tail feathers extending over the cushion.

miniatures of the large Cochins of the period. He wrote "Cochin bantams should be exactly the same as the larger Cochins, whilst as regard size and weight one fifth the weight of large Cochin fowls."

At this time many exhibition strains were developing a noticeable forward "tilt", with an outline not unlike a wedge of cheese. It was this type that was later favoured by British breeders, who regarded their birds as true bantams rather than miniature fowl.

For many years poultry exhibitions were dominated by large fowl. Most Pekin breeders had to exhibit their birds in the "variety" bantam classes. From the 1950s, shows were dominated by bantam entries and

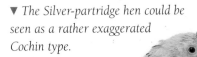

▼ The Silver-partridge hen could be seen as a rather exaggerated Cochin type.

▼ This Mottled hen is a good example of a British Pekin type and tilt.

▼ A Blue-mottled frizzled example has a fluffy outline.

▲ *Top row, left to right: Lavender hen, Silver-partridge hen, Lemon-cuckoo hen. Bottom row, left to right: Millefleur hen, Black male, Cuckoo hen.*

bantam exhibitors, with specialist shows attracting large numbers of Pekin bantams entries.

In appearance the Pekin is a bundle of plumage, with an abundance of soft underfeathers. The back has a significant rounded cushion of feathers. It has a characteristic forward tilt to its body with short legs and feathered

▼ *A Partridge hen.*

feet and toes. The breed makes a good broody and a good mother. They may be as tolerant of confinement as any true bantam but, given the run of a garden, a little group will adopt many of the sedate mannerisms of large Cochins.

ESSENTIAL CHARACTERISTICS

Size: Male 680g/24oz. Female 570g/20oz.
Varieties: Barred, Birchen, Black, Blue, Blue-mottled, Buff, Columbian, Cuckoo, Lavender, Mottled, Partridge, Red, Silver-partridge, Wheaten, White.
Temperament: Calm, gentle.
Environment: Copes with confinement. Requires grass on which to forage.
Egg yield: 60–130 small cream, tinted or brown eggs.

Sebright

The Sebright breed was named after its creator, who intended to produce an entirely ornamental bantam breed with laced feathering. It was the search for perfection in lacing, previously seen only in a rudimentary form in strains of the Poland breed, that led him to make an initial cross between the Poland and an unnamed "common" bantam. It is now thought that Hamburg and Rosecomb were also used in the genetic material of the breed. The offspring were bred and selected from until almost perfect lacing had been obtained and the crest of the Poland had been bred out. In spite of having Poland blood in their ancestry, the Sebright remains a typical true bantam.

In the Sebright breed the exhibition male should be "hen-feathered". This means that rather than having the curved and pointed sickle in the tail, and the narrow saddle hackle feathers that cover the back, each of these feathers is expected to be rounded like a hen's feathers, thus enabling every feather to be evenly laced. Not all males are hen-feathered, and those that hatch as such sometimes revert to standard feathering after a moult. Hen-feathered and standard-feathered males are both used in breeding programs by the more knowledgeable breeders.

After 200 years of inbreeding from a very limited gene pool, some strains of Sebrights have less than perfect immune systems. Additionally, some hen-feathered males are less fertile, making breeding Sebrights to reproduce to an exacting standard difficult,

▼ *A Gold-laced male with an upright posture and a bright and alert eye.*

and ensuring that exhibition breeds are likely to remain in the hands of skilled and dedicated enthusiasts.

Sebrights have distinct rich black lacing on a clear and even silver or gold ground. In total it took almost two decades to perfect the quality of lacing and to fix the breed type, though once it had been achieved, the breed quickly became a popular ornamental show bantam.

The breed has an upright and alert carriage, with a rounded breast carried forward and downward-pointing wings. The male has a rose comb. The skin is blue and the legs are clear of feathering.

▶ *A Silver-laced Sebright hen. In their early days, Sebrights were often referred to as Sebright Jungle fowl.*

ESSENTIAL CHARACTERISTICS

Size: Male 625g/22oz.
Female 570g/20oz.
Varieties: Gold-laced, Silver-laced.
Temperament: Alert.
Environment: An active and ornamental breed than can be quite hardy but is difficult to breed.
Egg yield: 50–80 small white eggs per year.

Serama

The Malaysian Serama evolved from a program of crosses between Japanese Bantams, Silkies and tiny Malaysian bantams known as Ayam Kapans. The genes of the Serama put it firmly in the true bantam camp. The breed was created in 1971 by the Malaysian Wee Yean Een, and named after the 16th century Thai king, Sri Ama. They are described as game-like because of their Malaysian bantam ancestry. The breed is so tiny that in its homeland it is often treated as a house pet, outnumbering cats and dogs. It is a good breed for urban areas, requiring just a small amount of living space. This breed of poultry is, in fact, the smallest in the world, and selection for size is a significant factor in breeding programs. In Malaysia it is exhibited in three different weight bands.

The breed is a relatively late export to the western world and is currently growing in popularity.

It has a pert and upright stance, a long, upright tail and equally long wings. In type the breed shows a "V"-shaped body posture, albeit in an exaggerated form, more than any of the other true bantams originating from the Far East.

As a relatively new breed, albeit distilled from ancient Malaysian true bantam populations, the Serama may take time to settle or be accepted into Western show culture. While they have been exhibited in a range of size bands, in Britain at least, the maximum weights seem likely to be adopted as the show standard for this tiny breed.

▲ *Spangled male.*

▲ *A White male with characteristic upright stance.*

▼ *A silkie Buff hen.*

▼ *A Millefleur hen.*

ESSENTIAL CHARACTERISTICS
Size: Male 500g/17½oz.
Female 300g/10½oz.
Varieties: No fixed colour varieties.
Temperament: Docile, friendly.
Environment: Said to require a very protected environment and to be difficult to breed.
Egg yield: White to dark brown.

A DIRECTORY OF MANMADE BREEDS

Manmade poultry breeds make up a distinct portion of all pure poultry breeds. These are the breeds that were created for commercial purposes by crossbreeding existing breeds, with the aim of producing offspring that benefit from the desirable traits of each parent, such as large body size and quantity of meat, their productive egg-laying capacity, or because they are economical to keep. Such breeds preceded the development of the modern hybrid, and for the most part were intended to provide maximum food for least cost. These fowl have been perfected over time by selecting and breeding from the best, and many have been used in the creation of new breeds. Today many of these breeds continue to have a commercial role, although most are not able to compete with the egg-laying capacity of hybrids. Like foundation breeds, those developed for exhibition purposes have less utility value, though the character and visual appearance of many attracts enthusiastic breeders and amateurs.

▲ *Burford Browns are good dual-purpose birds laying dark brown egg, but are not yet bred to an agreed standard.*

◄ *Lavender Orpingtons are a new variety of a well-known and popular manmade breed.*

WHAT IS A MANMADE BREED?

A manmade breed is a breed of poultry which has been developed by taking genetic material from at least two established pure breeds with the intention of creating a new breed that will inherit the most desirable qualities and characteristics of the parents.

The concept that a new poultry breed could be created out of two or more existing breeds was initially an American one. American breeders did not have any native chickens until immigrant settlers took fowl with them to the New World. The birds travelled as deck cargo to provide eggs and meat on the long sea journey. It is possibly because of this limitation that American breeders had far fewer inhibitions about making and naming new breeds of poultry from existing breeds than did breeders in other countries. In England, for instance, as late as the early 19th century the concept of a manmade breed was thought to be close to heresy. The idea that a breed could be created, stabilized and refined to an agreed standard that would then conform in certain external features such as size, comb, leg colour, and

type (shape), completely altered the way poultry keepers thought of poultry breeding. The driving motivation was purely commercial.

American breeders were prepared to use any breed in the breeding pen that could produce enough eggs in a

▲ *Marans are a French fowl developed for the very dark brown eggs that they lay.*

harsh climate to satisfy New York's ever-increasing demand for eggs. Unlike other parts of the Western world, since there was no local prejudice against yellow flesh, breeders could include an optimum proportion of Asiatic bloodlines in their table fowl breeding pens to produce the type of bird with plenty of breast meat that would satisfy market demand.

Developing breeds

The development of manmade poultry breeds followed two different, but parallel, courses. The first development was that of the "formula breeds", those where breeders included established breeds in their breeding pens with the aim of cross-breeding them to create new breeds

◄ *The extremely long tail of some Japanese breeds is a feature that has been selected in breeding programs.*

RHODE ISLAND RED x LIGHT SUSSEX: A MANMADE SUCCESS STORY

The British Light Sussex is an ancient foundation breed that evolved from management and selection of regional Surrey and Kent fowl. It has a high commercial value because it is a productive egg-layer. The Rhode Island Red is an American manmade breed created using genetic material from fowl imported from Europe and Asia. It, too, has a strong commercial value as a dual-purpose bird.

Rhode Island Red males were crossed with Light Sussex females to produce brown pullets and white males. The female offspring inherit their feather colour and egg-laying capacity from the Rhode male, and will produce up to 75 per cent of all eggs sold in Britain. The white or silver-coloured males inherit the Light Sussex coloration and enough of her body conformation to make them a worthwhile table fowl.

This feather-colour sex-linkage allows early visual determination of chick sex,

and this cross is the basis of much of the world's brown egg-laying hybrid flock. This egg production gene is carried down the male line. Added to that, the Light Sussex is a very productive hen. The two breeds crossed together therefore produce a high egg yield. The more commercial strains of the Rhode Island

▲ *The Light Sussex (left) and Rhode Island Red is a classic pairing with offspring that are prolific egg-layers.*

Red family are some of the most productive hybrid strains and they are one of the success stories of modern agricultural development.

encompassing the best features and character traits of each breed. Good examples of type were required for the original breed stock. The second course saw new breeds such as the Rhode Island Red gradually evolve out of a mixed population of fowl from all over the world, rather than from specific breeds. Using genetic material from a greater pool, it became possible to perfect the traits of the offspring. The new breeds were developed to adapt to local conditions as well as to meet market demands. The breeding program for each type is rigorous, with the requirement that the resultant offspring are stable and will breed true to type, if required. When it became generally accepted that the standard breeding formulae could be altered to create new colours, poultry breeding and

exhibiting became a more creative pastime, attracting fanciers with an interest in perfecting colours. At that time poultry farmers understood the concepts of varieties within a breed, the various strains and the importance of utility selection, but at that period emphasis was on breed type and breed standard, rather than on hybridizing.

Lasting legacy

The implications of rearranging breed formulae to create new colours, as well as bantam counterparts of most breeds, have been considerable. Often, many manmade bantams are now more popular than the original large breeds. The ethos behind the creation of many of these miniatures was to recreate, as closely as possible the large breed characteristics. When

bantam versions of large breeds were created, the wings were held far higher and at a more horizontal angle.

Crosses of Rhode-type males with hens descended from and not dissimilar to the original Light Sussex are the basis of many of today's black-tailed brown hybrid hens. A similar "recessive white" female line, descending in part from single-combed sports of White Wyandottes and white sports from Rhode Island Reds, is crossed with a typical Rhode Island Red male to produce the common pink- or white-tailed brown hybrid hens. Many of the almost black hybrids that are popular with small producers are the product of crosses involving Rhode Island Reds and Barred Plymouth Rocks. The significant manmade breeds still influence modern poultry breeding.

AMERICAN MANMADE BREEDS

American manmade breeds are the results of experiments which took place probably while the new nation was forming. With no indigenous chicken breeds in North America, American breeders had few qualms about creating fowl, that combined the best characteristics of existing breeds. In the early days, such breeders were driven by purely commercial interests. Farmers and breeders selected from a wide variety of breeds to create new dual-purpose types that would be prolific egg-layers and also satisfy the growing demand for meat caused by rapid urban expansion. Unlike European consumers, no prejudice against yellow poultry meat was felt. This meant that breeders were free to use Asian fowl in their programs. Crosses between Asian, European, Mediterranean and British fowl yielded breeds such as the Rhode Island Red, today probably the world's most successful poultry breed. The first American poultry show was held in 1849. The American Poultry Association was founded in 1873, arising from a need to set standards for poultry breeds and to appoint judges. Only one year later, the first *American Standard of Perfection* was published, and it remains one of the best-respected poultry breed handbooks in use today.

Dominique

The origins of the Dominique probably represent one of the earliest examples of the fusion of the ultra-light Northern European breeds with one of the massive Asiatic breeds. Unlike the later, similarly marked and formula-bred Barred Plymouth Rocks, no written account of how Dominiques were created exists. This suggests that similar fowl occurred whenever black Asiatic fowl were introduced to an area where Pencilled Hamburg-type fowl had a predominant influence on the local fowl population. The Dominique has the Hamburg's rose comb and yellow legs. The lack of foot feather suggests that the breed has clean-legged Java-type fowl in its ancestry. The breed has significant amounts of feathering. Such breeds had existed in the USA since 1835 and would have figured more prominently than any of the Cochin or Shanghai alternatives in any breeding program. However,

> ESSENTIAL CHARACTERISTICS
> **Size:** Male 3.2kg/7lbs.
> Female 2.3kg/5lbs. If bantams existed they would be expected to weigh 20–25 per cent of the large versions.
> **Varieties:** One only.
> **Temperament:** Reasonably quiet.
> **Environment:** Tolerates confinement. Likes to forage. Cold hardy.
> **Egg yield:** Up to 230 medium-size brown eggs per year.

the addition of these genes could have resulted in the breed's red face.

Single-combed Dominique-type fowl are thought to have played a part in the later creation of Barred Plymouth Rocks. This suggests that fowl type may have been well known among poultry breeders. These early crosses would have inherent hardiness from their north European ancestry, and size and productive egg-laying properties from the Asiatic family.

Dominiques are an interesting, active, garden-worthy breed, and are good dual-purpose birds. Hens are known to go broody and make good mothers; the sex of the chicks can be determined upon hatching. This breed was never reared commercially.

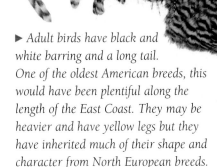

▶ *Adult birds have black and white barring and a long tail. One of the oldest American breeds, this would have been plentiful along the length of the East Coast. They may be heavier and have yellow legs but they have inherited much of their shape and character from North European breeds.*

Jersey Giant

As its name suggests, the Jersey Giant is a large fowl that originated in the US state of New Jersey. When first bred in the 1870s, it would have evolved to help meet the increasing demand for large, well-fleshed table fowl, as called for by the New York and Philadelphia markets. Unlike the British breeders, whose market was prejudiced against fowl with yellow legs and yellow flesh, and thus were inhibited from using large proportions of Asiatic genes in their breeding programs. American poultry breeders could maximize their use of Asian breeds with yellow flesh, such as Brahmas and Javas in their breeding programs, as American consumers had no prejudice about skin or flesh colour.

The breed was admitted to the American standard of perfection in 1922 as Jersey Black Giants, after the name of the original breeders, John and Thomas Black. Black was the original colour of the fowl, although later Whites were produced and then Blues, with all varieties possessing legs with an underlying yellow or willow pigmentation. Over the years, however, the term Giant has been called into question by those comparing the size of the breed with large fluffy breeds such as the Orpington, which have been bred to an even larger size.

ESSENTIAL CHARACTERISTICS

Size: Male 5.9kg/13lbs.
Female 5kg/11lbs.
Varieties: Black, Blue, White.
Temperament: Extremely quiet, slow-moving.
Environment: Large housing required.
Egg yield: 135–150 eggs per year.

▼ *The Jersey Giant is a massive bird, but has been surpassed by other breeds in terms of size since it was developed.*

The breed's creators used Black Java, Cornish Game, Dark Brahma and Black Langshan in its genes. When they did so, they would have been seeking to create a breed with not just size and weight, but weight in the right proportions and on the areas of the carcass that make a good roasting fowl, with the intention of marketing the bird as table fowl. This was a bird that was intended to compete with the turkey as a roasting fowl of quality. The original Jersey Giants would also have been made or have evolved to fill a gap which had formed when breeds such as the Rhode Island Red were developed as egg-laying breeds rather than table fowl. The Jersey Giant has remained a good enough egg-layer to produce enough chicks to meet this demand for table fowl. The breed has never been sold commercially as an egg-laying breed, however, since the ratio of food it eats to the number of eggs it produces makes it non-viable. Conservation of the breed is left to the enthusiastic amateur.

Jersey Giants are friendly and docile, and the males are rarely aggressive. They adapt well to both confinement and free ranging. Housing is required for roosters of up to 5.9kg/13lbs as well as room for hens up to 5kg/11lbs in weight. As a result of their size, this breed requires a reasonable amount of exercise from an early age for the birds to develop the legs, bone structure and musculature needed to carry the sort of weight expected of a breed labelled "giant". Perches will need to be placed reasonably close to floor level, and where adult birds are found reluctant to perch, any sleeping or resting area will need to be covered with clean wood shavings or similar material. When held, they are "lumpy". The breed has a well-developed breast and carries considerable amounts of meat which are distributed across its wide, flat back. There would be little point in creating a bantam version, because it might well be bigger than the large versions of many other breeds.

Plymouth Rock

A dual-purpose fowl, the Plymouth Rock is a highly successful breed of poultry. The Barred variety saw American poultry breeders embark on a new chapter in poultry breeding. A program of crossbreeding fowl created hybrid offspring with beneficial vigour. Breeders hoped to encapsulate this trait, along with the benefits of the parents' pure and identifiable breeds in their offspring.

Plymouth Rock No 1 was created from a three-way cross between a Black Java, an unidentified Asiatic hen and a barred Dominique. The offspring from this cross was bred with a White Brahma.

Plymouth Rock No 2 was the result of a cross between a Dominique male and a Black Java female. The offspring of this union were feather-legged, and played little part in any later breeding program.

▼ *A Partridge Plymouth Rock bantam hen.*

A Minorca was mated with a Cochin cross, the offspring of which was back-crossed to Plymouth Rock No 1 to produce Plymouth Rock No 3. In turn, this was mated with a Dark Brahma to produce Plymouth Rock No 4. The final step was a back-cross between Rock numbers I and 4 to produce the first Barred Plymouth Rock.

While the type (shape) and colour of the Plymouth Rock would have been established with these preliminary crosses, the complex work of selecting strains that combined all the desirable qualities of their ancestors within the package

▶ *The Barred Plymouth Rock is one of the most popular colours for this breed, which was developed as a dual-purpose bird.*

of a single strain was only just beginning. Initially, these first Rocks would have enjoyed enormous amounts of hybrid vigour along with the potential to produce the odd rogue bird that carried less useful traits, which then had to be eliminated from the breeding program. Contemporary American sources refer to the emergence in 1878 of an Essex strain, which, like most strains, either merged with other strains, became too inbred or had to be out-crossed to lose the breed's original genetic intensity.

Over the years, every really successful breed has relied for its creation on countless, often inbred, strains that could later be crossed to create new strains. In the case of the large Plymouth Rocks, there have been strains developed as table fowl and others selected for an ability to lay well, as well as to excel as parents. Most of the black hybrids favoured by many smaller-scale poultry keepers contain at least 50 per cent utility Barred Rock in their ancestry. More controversially, the search for the

ESSENTIAL CHARACTERISTICS
Size: Male 4.3kg/9½lbs.
Female 3.4kg/7½lbs. Bantam sizes
should not exceed 1.3kg/3lbs for the
male and 1.2kg/2½lbs for the female.
Varieties: Barred, Blue, Buff,
Columbian, Partridge, Silver-pencilled,
White.
Temperament: Quiet.
Environment: Free range or semi-
intensive. Resistant to cold.
Egg yield: Exhibition strains may lay
120–180 eggs. Utility strains may lay
more than 200 light brown eggs.

finest possible barring saw some exhibition strains with very narrow feathers, bred to such an extent that the overall vigour of the more exaggerated examples was lost. While the Buff Plymouth Rock, created from Buff Leghorns, Buff Cochins and Light Brahmas, had a completely different ancestry, the development of strains, each modelled on the Plymouth Rock type, saw them quickly accepted as a variety of the same breed.

The Barred version of the Plymouth Rock probably represents one of the first successful attempts at using a formula to create a new breed containing the most desirable characteristics of its parent breeds. White sports that appeared among some strains and families of Plymouth Rock were back-crossed to form emerging strains of Buff Rocks. Later, some of the heavier white sports, sometimes known as Albions, were used in the complex mix that went into the creation of the later broiler strains.

Both the Partridge and the Silver-pencilled varieties would have had an infusion of similarly coloured Wyandotte blood. Many unusual varieties are likely to remain scarce

particularly in large fowl; they are often more abundant as miniatures.

The same format that allowed breeders to create new colour varieties enabled miniature versions to be developed, with the Barred and Buffs becoming some of the most popular exhibition varieties. The Partridge variety, which is one of the most attractive as well as one of the hardest to perfect, is now found in a form that can rival any Wyandotte in terms of egg production. With feather-free legs, these bantam versions are reasonable layers and a most attractive and garden-worthy miniature fowl.

In appearance this dual-purpose bird is large, with a long and broad back, a

▶ *Plymouth Rocks are one of the world's best-known and popular poultry breeds.*
This is a Partridge rooster.

▲ *A Black Rock is a commercial hybrid developed from the Plymouth Rock breed. It has been a successful breed, readily producing a good egg yield.*

moderately deep breast, and yellow legs, beak and skin. The hens possess a deep abdomen, which signifies a productive egg-layer. The face is red, with red earlobes and a single comb.

This is a hardy and long-lived breed. In nature it is docile and the hens make good mothers. The hen has been most useful in producing sex-linked hybrid crosses, meaning that the sex of the offspring can be identified by the colour of the chick's down on the first day of its life. The most famous hybrid that the Plymouth Rock hen has produced is the Black Rock. Fine exhibition examples exist, but selecting birds for the quality of their barring tends to be at the expense of the egg-laying gene.

Wyandotte

While utility considerations seem to be the driving force behind the creation of most breeds of poultry, rather more exotic aims drove the focus of attention that eventually led to the Silver-laced Wyandotte, and later to the whole family of Wyandotte varieties. It seems that the original intention was to improve Cochin/Pekin bantams by crossing a male Sebright bantam with a Cochin hen. When the result turned out to be too big for a bantam, the offspring were sold as Sebright Cochins. A second cross between a Silver-spangled Hamburg and Buff Cochin was made. The offspring of these two crosses, when mated together, resulted in American Sebrights. However, without any agreement on comb type, admittance to the American Standard was refused in 1876. At about the same time, a cross between a Silver-spangled Hamburg and a Dark Brahma hen produced a more desirable pea comb and was given

▲ *The original colourway is the Silver-laced variety, shown here on a large hen.*

ESSENTIAL CHARACTERISTICS
Size: Large male 3.9kg/8½lbs.
Large female 2.7kg/6lbs. Bantam weight not to exceed: male 1.7kg/3¾lbs; female 1.4kg/3lbs.
Varieties: 17 known including Black, Blue, Blue-laced, Buff, Buff-laced, Columbian, Lavender, Gold-laced, Partridge, Red, Silver-laced, Silver-pencilled, White.
Temperament: Docile, friendly.
Environment: Cold hardy, prefers a free-range environment.
Egg yield: Exhibition strains lay 70–160 eggs per year. Utility strains of White lay 200–250.

the provisional breed name Eurekas. The original ancestry of the Wyandottes included Pekins, Cochins and Silver Sebright bantams. The resulting breed has been endowed with a mixture of all the necessary genes used to develop some of the most spectacular

exhibition fowl, as well as some of the most productive utility strains. When these were bred with the offspring of an earlier Silver-spangled Hamburg-Dark Brahma cross, the result was a uniform rose comb that became an essential part of the breed standard. They were admitted to the American Standard as Wyandottes in 1883.

Given that Buff Cochin went into the original mix, one could have expected the odd Gold-laced example to turn up as a sport. However, the breeder Joseph McKeen of Wisconsin claimed to have first produced a Winnebago breed from a Partridge

▶ *A Partridge bantam hen. It is the quality of lacing and the even distribution of lacing over the body that is the feature most difficult to perfect.*

▼ *A large Lavender hen.*

▲ *A Buff-laced bantam hen.*

▼ *A Blue-laced bantam male.*

Cochin, a rose comb Brown Leghorn and a Gold-laced Sebright. The Winnebago was crossed with a Silver Wyandotte to make the Gold-laced Wyandotte.

The Cochin would have given the breed its cushion tail, its curvaceous outline, and the rose comb with neatly downturned leader. The Silver-pencilled variety could owe much to the original Dark Brahma genes, and the Partridge to the Partridge Cochin.

Even without further genetic input, all the elements were now in place for the breed, with selection, to fulfil almost any role. The Silver- and Gold-laced varieties that arrived in Britain in the 1890s were soon established as good egg layers. The Whites and Blacks, as sports, had been noted for laying during the winter months (always a desirable characteristic), and strains of Whites had broken egg-laying records.

Bantam versions, probably resulting from the original true bantam Sebright input, remain exact replicas of the original stock. Unfortunately, the faults inherent in the breeds used in the original formulae can still manifest themselves in the offspring; parents with perfect combs can produce chicks with single ones.

Many miniature breeds are on the large size for bantams. Most Whites have the capacity for excess feather and fluff that, while welcomed by most exhibitors, can lead to the more exaggerated examples being extremely unproductive.

Some of the more intricately marked examples are often found to lay well. The breed is a quiet domestic fowl that is content with less space than some other large exhibition fowl.

▼ *A large White male.*

▼ *A large White hen.*

▼ *A Partridge hen.*

New Hampshire Red

The early New Hampshire Reds had the same genetic basis as the Rhode Island Reds, which in turn owe much to the work of the Rhode Island Experimental Institute. New Hampshire Reds are made solely of Rhode Island Reds, but have been selectively bred to create a bird with different strengths. Work done at the University of New Hampshire first encouraged the selection of the New Hampshire Red as a separate breed. It was the vision of poultryman A. W. (Red) Richardson, who recognized that mountain-bred poultry stock had sterling properties, and encouraged scattered breeders to collaborate with a common aim. In 1935, the breed was standardized in a natural orange-brown colour with limited Columbian-black markings.

▼ *A miniature hen.*

The breed made little impact when first imported into Britain in 1937, but that changed in the 1950s when it found its way into the hands of a long-established Sussex breeder, who discovered that the breed made better cross-breeding fowl than Rhode Island Reds. The first results were too unremarkable to attract anyone other than those looking for the most commercially viable breeds, but shortage of bloodlines saw the importation of much rounder and more feathery strains, first from Germany, and then from Holland. The result was a much rounder, more comfortable-looking fowl that had very attractive coloration. It attracted enough interest for breeders to form their own breed club. In its present form, the New Hampshire Red averages nearly 200 eggs a year. Unlike the Rhode Island Red, breeders were intent on producing a bird that matured quickly and had good meat quality. As a result this breed is now a less productive egg-layer than its ancestors.

In appearance the New Hampshire Red is similar to the Rhode Island Red, but with lighter coloration. It has a single comb, red wattles and lobes, and is an easy-going fowl that is close to being a perfect domestic breed because of its dual-purpose characteristics.

The bantam versions may be even better egg-layers than the large variety, and are beginning to gain recognition on the show circuit.

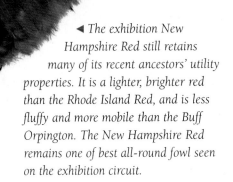

◄ *The exhibition New Hampshire Red still retains many of its recent ancestors' utility properties. It is a lighter, brighter red than the Rhode Island Red, and is less fluffy and more mobile than the Buff Orpington. The New Hampshire Red remains one of best all-round fowl seen on the exhibition circuit.*

ESSENTIAL CHARACTERISTICS
Size: Male 3.9kg/8½lbs.
Female 2.9kg/6½lbs.
Varieties: One only.
Temperament: Placid nature, good for a garden environment.
Environment: Tolerates confinement.
Egg yield: 200 brown eggs per year.

Buckeye

The American Buckeye is a dual-purpose breed with lustrous red plumage. The breed was first bred in Ohio and was developed in the last decade of the 19th century by Mrs Nettie Metcalf of Warren, Ohio, who crossed a Buff Cochin with a Barred Plymouth Rock hen. The female offspring were crossed with black-red game males, and the resulting generation formed the basis of selection for the new breed.

The Buckeye breed has a unique body shape, although the breeder was aiming to replicate the Indian/Cornish Game breed. It is reported, however, that the Cornish breed was not included in the genetic composition of the new breed.

The body is broad and squat, with a back that slants towards the ground. As a dual-purpose bird, the breed was developed to have strong, meaty thighs, and well-developed wings and breast. In colour it is deep red, which may cause confusion with the Rhode Island Red breed. However, a degree of the Buckeye retains the dark-coloured barring of its original parent in the soft downy underfeathers on the back. Rhode Island Red feathers, in contrast, are uniformly red. The feathers are tightly packed – a trait that characterizes all birds in the American class. The Buckeye has yellow legs and skin, red lobes and comb and a pale brown beak. Its pea comb signals that this is a hardy breed and can tolerate cold winters.

There was talk of calling the Buckeye a Pea-combed Rhode Island Red, but it quickly became apparent that the name would be detrimental to the breed's popularity. Instead, the name Buckeye was chosen in honour of the state in which it was bred. This breed was developed prior to the Rhode Island Red, and it is possible that the Buckeye was used to improve the genetic base of the Rhode Island Red. The new breed was admitted to the American Standard in 1904. However, it is now regarded as critically endangered in the USA.

The breed is an active bird, which is friendly towards humans, and so makes a good family pet. It will tolerate a variety of environments but prefers a free-range setting where it is able to forage for food. It does not particularly thrive in close confinement.

▼ *A large hen with characteristic rounded chest and substantial, solid appearance. With a diminished gene pool, some, like this example, are found with less than perfect pea combs.*

▲ *A large male. The tail feathers of the male are particularly long.*

ESSENTIAL CHARACTERISTICS

Size: Male 4.1kg/9lbs.
Female 2.9kg/6½lbs.
Varieties: One only.
Temperament: Friendly, active.
Environment: Dislikes confinement, tolerates cold.
Egg yield: 150–200 medium-size brown eggs per year.

Rhode Island Red

Early breeders of the Rhode Island Red – the poultry farmers of New England and Rhode Island – were motivated to produce poultry that would meet the increased commercial demand for eggs and poultry meat made by the expanding local urban population. Unlike breeders of other poultry breeds, their interest was focused on producing a bird that met the requirements of their customers rather than to comply with a written standard for their breed. It would have been the Rhode Island Red's early development as a dual-purpose fowl, that led to it becoming the world's most successful breed.

For most of the 20th century, the Rhode Island Red was the most important brown egg-laying breed of poultry. One of the first farmers and market traders to start creating a purpose-bred strain was William Tripp, who, in 1854, obtained a large black-red Malay rooster that had arrived from a South-east Asian port. He bred this with his scrub hens and noticed that the resultant chicks were far superior to other local fowl. The

▲ *A rose comb version of the Rhode Island Red.*

birds produced better meat and the hens laid more and bigger eggs. His friend, John Macomber, who lived in Westport, Massachusetts, became interested and the two

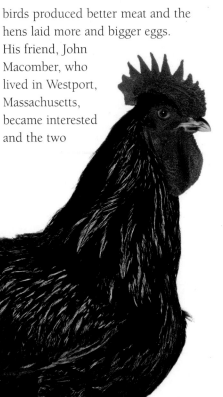

▶ *The Rhode Island Red has a docile, friendly and quiet temperament, and as such, makes a good family pet. However, males are known to attack strangers.*

▲ *The single comb of the Rhode Island Red male stands upright.*

worked together for a time crossing and exchanging their best birds to avoid too much inbreeding. After Macomber died, the work was taken up by others who sought to breed fowl that would produce bigger and browner eggs while at the same time providing a larger proportion of meat. Malay, Cochin, Langshan, Hamburg and Leghorn genes are all thought to be included in its genetic composition. The aim was to produce the hardiest stock that would prosper under any conditions, including the rigorous New England winter.

Much of the work to stabilize the breed was done by the Rhode Island Experimental Station, which selected and collected birds from a wide area. When Rhode Island Reds were first seen in Britain, they were reported as being pea-combed, even though many of the early imports into the United Kingdom were described as rose-combed. In the USA the single comb was admitted to the standard in 1904. The rose comb was standardized as Red American in 1905, but by 1906 the American Poultry Association recognized both combs in the breed, as did the British Rhode Island Red Club when it was formed in 1909.

ESSENTIAL CHARACTERISTICS

Size: Male 3.9kg/8½lbs.
Female 3.4kg/7½lbs.
Varieties: Red.
Temperament: Good family pet, though can be aggressive to strangers.
Environment: Likes to free range, and is cold hardy. The comb can become susceptible to damage in extreme cold.
Egg yield: At least 200 brown eggs per year, though, like many breeds, productivity drops in very cold temperatures.

▼ *Rhode Island Red bantam hens like this one are highly productive egg-layers, including those birds bred for exhibition purposes.*

While some British breeders would have selected strains to meet their own interpretation of the standard, the original wording remained the basis of birds selected to be utility and commercial laying fowl. When entered in laying trials, the foremost utility strains would often lay more than 280 eggs, making this breed one of the best for farmers, smallholders, and domestic keepers. While most of the large Rhode Island Reds seen at shows today may be too big and not lay enough eggs to be described as utility, the exhibition-bred bantams are excellent layers of reasonably sized brown eggs. For those who still want to keep useful representatives of the best-known historic fowl, some of the old commercial stock is still kept, including one known strain that has remained as a closed flock since winning a National Laying Trial gold medal in 1954.

The Rhode Island Red has rust-coloured plumage that can be as dark as mahogany or significantly paler; the male can have black on the wings. The breed has a pale brown beak and pale yellow legs. With its rectangular shape, broad, flat back and medium-length tail, this fowl is often described as being brick-shaped. It is a medium-weight poultry breed and makes a good family pet. It is especially suitable for owners interested in egg production, although strains bred for exhibition are notably less productive than those that have retained their utility value. By nature the breed is docile and friendly, particularly to those responsible for feeding it, though it can be aggressive to strangers. It is an alert bird that likes to forage and free range. These birds often remain together as a group.

The breed fell out of favour with those keeping hens for egg production in the 1970s, when hybrid birds were proved to lay greater quantities of eggs for less feed than standard breeds. However, the breed is still the best known in the world, and is often used as the parent of many of today's hybrids. The male, in particular, is considered useful since the egg-laying gene is passed down the male line.

► *Rhode Island Reds make good, if large, pets that add a valuable contribution to the garden when allowed to free range.*

Araucana

The Araucana breed is considered to be South America's only native fowl. Its most distinguishing feature is its blue eggs. Throughout the world there are many different types of fowl, all with differing visual characteristics, that are capable of producing blue eggs, and all are known as Araucanas. This is because the blue egg-laying trait or gene is more dominant than most external breed characteristics and, for this reason, the trait is used to determine the breed type. Two are native to Chile: the ear-tufted Quetros breed and the rumpless Colloncas breed. In 1914 it was suggested that the blue egg-laying gene could come from a wild fowl, the Chachalaca breed, which had reportedly hybridized with domestic fowl in Chile.

A number of forms of the Araucana are found in the USA, though a rumpless version with unique ear tufts on the ends of unique fleshy or gristly ear appendages is generally thought of as being the true descendant of the Chilean Araucana. This descendant, like many original forms of other ancient fowl populations, is not large enough to be generally accepted as a large type or as small as most other bantams. More unfortunately, the gene that seems to be linked to ear tufts also results in varying proportions of chicks dying

before they hatch. Rumplessness also gives rise to fertility problems, so if these versions are the purest form of Araucana, they seem destined to remain in the hands of a small group of enthusiasts.

Versions of the breed with a tail are generally free of inherited weaknesses and will be the best choice for those attracted to rearing breeds that lay blue eggs. Those standardized as British Araucanas have a small head tuft. A similar tuftless breed is standardized in the USA as Ameracauna. The most distinguishing feature of the breed, the blue egg, is the reason for its popularity; blue eggs sell well in farmers' markets, and their saleability has encouraged the poultry industry to attempt to produce enough similar blue eggs to supply the commercial market. However, like most small-scale poultry keepers before them, the industry has found that the Araucana and its descendants are remarkably resistant to laying large quantities of very large eggs. Strains of the standard breed are available that lay fairly large numbers of mid-size eggs; in order to increase the egg yield, the breed has been crossed with other breeds. First-generation offspring resulting from outcrosses with white egg-laying breeds lay pale blue eggs. The same crossing with brown egg-laying breeds produce olive or greenish coloured eggs. For those interested in exhibiting eggs, a large deep blue egg will take many prizes. The bantam version, which can be quite small, may also lay lots of pretty blue eggs. The breed's cross-bred form has been

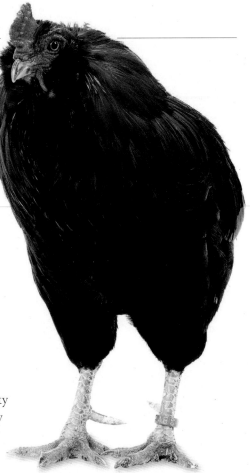

▲ *The blue egg-laying Araucana remains the basis of nearly all of the fowl that lay the majority of blue eggs.*

used to create commercial breeds. The inclusion of the Araucana breed in the Crested or Cream Legbar had a lasting commercial impact.

▶ *Large and bantam versions are standardized in a wide range of colours.*

ESSENTIAL CHARACTERISTICS
Size: Male 2.7–3.2kg/6–7lbs.
Female: 2.25–2.7kg/5–6lbs.
Varieties: Black, Black-red, Blue, Crele, Lavender, Gold-duckwing, Pile, Silver-duckwing, Spangled, White.
Temperament: Alert, active, hardy.
Environment: Garden.
Egg yield: 270 green or blue eggs.

EUROPEAN MANMADE BREEDS

European poultry breeders have always viewed the utility and commercial aspects of poultry breeding in a rather different light from their British and American counterparts. The French saw the quality of their regional table fowl as being of primary importance. They had long thought that quality of meat far outweighed considerations of poultry size. When creating the Faverolles and Houdan breeds, the British Dorking was also included in the genetic composition; a sign that French breeders recognized the quality of its meat. As a result both breeds had a fifth toe like the Dorking. Both the French and Dutch, who each had a surprisingly large share of the British egg market, later developed breeds to meet the British preference for the darkest brown egg in the hope of selling into that market. The fact that the market was willing to pay a premium for dark brown eggs influenced the development of both the Dutch Welsummer and Barnevelder, and the French Marans breeds.

Houdan

The Houdan is an old French breed named after the town of that name which is situated 40 miles from Paris. The town of Houdan was where the important poultry markets which supplied the Paris meat markets were located. The breed's origins can be traced back to the start of the 18th century. The Houdan was originally raised as a dual-purpose bird. It is heavy enough to produce a meaty carcass and its meat was and still is considered to be that of a quality table fowl with

▲ *The breed has an interesting "leaf" comb and pretty mottled feathers.*

succulent white flesh. It also lays approximately 160 small- to medium-sized white eggs per year, converting feed to eggs at a very economical rate. In its ancestry, it is thought to have Poland and Crèvecoeur, which gave the breed its crest, and Dorking, which gave the breed its five toes. Most poultry have four toes. It is now regarded as an ornamental fowl, and can be particularly spectacular to look at with its attractive crest, beard and muffs. It has small ear lobes and wattles, which remain mostly hidden by the head plumage. Many breeders concentrated on the ornamental points of the Houdan, at the expense of its utility value. The Houdan is a docile breed that can cope well with confinement, although care needs to be taken to protect the crest. It is an excellent breed to keep in the garden and will happily forage for food. Total free-range conditions are not always desirable, particularly for the rooster, since his crest can obscure his vision making it difficult for him to see predators and move out of the way.

◄ *The Houdan is an attractive bird that makes a good family pet. If handled from a young age, it responds well to being petted.*

ESSENTIAL CHARACTERISTICS

Size: Large male 3.1–3.6kg/7–8lbs.
Large female 2.7–3.1kg/6–7lbs.
Bantam male 680–793g/24–28oz.
Bantam female 623–737g/22–26oz.
Varieties: Mottled, White.
Temperament: Quiet if handled young.
Environment: Free range, but crest needs protection from rain.
Egg yield: 160 small to medium white eggs per year.

Barnevelders

The Barnevelder breed evolved from landrace poultry that were kept in the Barneveld district of the Gelderland province in Holland. In 1930, a World Poultry Congress report stated that the breed origins were those of local farm fowl crossed with Cochin hens. The offspring of these hens were crossed with Brahmas and later with Langshans to produce the bird that we know today. In 1906, the breed was crossed with Buff Orpington and in consequence its colouring changed to that of Partridge. However, the breed standard had been established long before this report. A photograph of a pen containing four hens and a male, which were winners at the Utrecht, Netherlands, show of 1922, shows a foppish breed with the same pronounced U-shaped back that forms part of the breed's standard.

The popularity of the breed was due to the desirable brown eggs, which were in great demand. The breed's standard was at one time prefaced with the note: "The egg is part of the standard, no more no less". The coffee-brown egg is mainly due to the Langshan genes, although the egg of the present Barnevelder is larger. The breed also remains productive throughout the winter.

▼ Crisp double lacing on a rich red-brown ground makes the Barnevelder hen one of the most attractive utility fowls.

ESSENTIAL CHARACTERISTICS

Size: Large male 2.7–3.6kg/6–8lbs.
Large female 2.3–3.1kg/5–7lbs.
Bantam male 680–793g/24–28oz.
Bantam female 566–680g/20–24oz
Varieties: Black, Double-laced, Partridge, Silver, White.
Temperament: Quiet, docile.
Environment: Small runs and free range.
Egg yield: 160–200 large brown eggs per year.

Breeders kept Barnevelders firstly for utility value and secondly for feather colour. Early reports stated that the breed laid very dark brown eggs. Today, however, Welsummers and Marans both lay darker brown eggs than the original Barnevelders ever did. Show breeders now concentrate on visual features rather than on egg colour. Originally it was egg size that seemed to be a feature of their participation in laying trials. One pullet from the Bethnal Green Utility Poultry Society laid 201 "specials", 39 "firsts" and 1 "second" when entered in the Middlesex trial of 1934. In build the Barnevelder is medium-size, producing a good carcass of meat.

Both Partridge and Double-laced varieties were bred in Britain. The Laced variety soon established itself as one of the most attractive colour patterns that is still compatible with a utility breeding program. Partridge examples are still seen on occasions: some birds of this strain give the appearance of having been bred from a laced ancestor. It is unlikely that a breed that, in its large form, at least, has a relatively small gene pool, can fully support two complex colour varieties while maintaining the breed's other important characteristics. Blacks that used to be bred in some numbers still turn up as sports from the double-laced variety, probably a throwback to the earlier Langshan ancestry, but this would probably be very difficult to maintain in both sexes with the required yellow legs. This is a neat and compact bird with a very upright stance and a single comb.

Few pure breeds are now kept commercially. The Barnevelder, in both large and bantam forms, remains one of the most attractive breeds that could still, with careful selection, pay its way in a domestic situation.

Welsummer

Like most of Europe's poultry breeds, nearly all of Holland's indigenous fowl would have laid white or tinted eggs. Dutch poultry breeders were among the first to recognize the commercial importance of dark brown eggs, which were perceived to be a healthier choice than white eggs. The Dutch Welsummer breed is famed for its dark brown eggs. The main source of brown pigmentation in the shell is derived from the heavy Asiatic fowl that were imported into mainland Europe during the 19th century. There also appears to be a link between the U-shaped back outline of the Asian Langshan breed and the dark egg colour of the Welsummer.

The original fowl that were the foundation of the Welsummer breed were from an area north of the river Ysel. They were reported as being of very mixed colours and types, with some even having five toes. By 1917, breeders had established some breed uniformity, with only red cockerels with the markings of the partridge-coloured breeds being used in breeding pens. It is likely that crossing this

landrace breed with Barnevelders also created colour stability, although Brahma, Cochin, Wyandotte, Leghorn and Rhode Island Red could also have added to the gene pool.

The Welsummer is a large, upright bird with a deep breast, broad back and large tail. It carries a single red comb, large wattles and has yellow legs. This is the breed that appears on the famous cereal packet. Welsummers are active, friendly and placid birds that suit a garden environment. Bantam versions are available, but they produce a lighter tinted egg.

By 1930 the British Welsummer Club was formed, and from the outset there seems to have been much rivalry between Dutch and British breeders and considerable controversy over breed standards. Since the breed was imported to the UK before it was fully established, this left plenty of scope for the breed to develop in parallel in Holland and in the UK. Not only did those who first imported birds from Holland hope to corner some of the market for dark brown eggs, they also hoped to make money by supplying breeding stock to other poultry keepers at a time when there were no restrictions on importing birds. In Britain, breed development has focused on the competitive

◀ A Partridge bantam pullet.

element of exhibiting the dark brown eggs. Without the same stimulus of egg shows, the breed as now kept in Holland lays only mid-brown eggs.

Most small-scale poultry keepers would be happy to find a strain of large Welsummers that lay 180 dark, matt brown or "flower-pot red" eggs, each weighing 70g/2½oz. However, the darkest eggs are laid by pullets which produce fewer eggs. It makes no sense commercially to have a hen that produces dark brown eggs unless it lays a reasonable number. The breed is standardized as light but is weightier than some heavy breeds.

▶ Described in Holland as being Rust Partridge, the female Welsummer is enhanced by a rich chestnut-brown breast and golden-brown neck hackles.

ESSENTIAL CHARACTERISTICS

Size: Large male 2.7–3.1kg/6–7lbs. Large female 2.2–2.7kg/5–6lbs. Bantam male 1kg/36oz. Bantam female 793g/28oz.

Varieties: Partridge, Silver-duckwing.

Temperament: Active, but quiet.

Environment: Free range or grassed runs.

Egg yield: 160–200 dark brown spotted large eggs per year.

North Holland Blue

The Assendelft breed of poultry had been kept for centuries in an area north of Amsterdam, and were considered typical of the small fowl of northern Europe. This breed was tolerant of wet land, but laid small eggs. When breeders looked to improve the egg size of the breed, the Belgian Maline breed provided the perfect genetic qualities to introduce to the breeding pen. The Malines had a similar tolerance to wet ground, but were distinguishable from the Assendelft by their leg feathering. This breed was also used in the genetic composition of the Marans, and would have been responsible for adding greater weight to both breeds.

Standardized as North Holland Blue more than 100 years ago, some of the first birds that arrived in Britain still had feathers on their legs. Poultry judges of the time decided that the breed would be indistinguishable from Marans unless it was devoid of leg feathers. Leading utility judge W. Powell Owen persuaded the British Poultry Club to standardize this Dutch breed as having feathers on its legs. However, when British poultry farmers imported large numbers of chicks from Holland, they found that nearly all of these were clean-legged, and with the British table market in mind, they continued to select for clean legs. As long as the breed remained on poultry farms, those wanting to exhibit their birds were likely to find a farmer willing to part with the occasional bird that had feathers on its legs for them to use as breeding material.

In many ways, these imported North Holland Blues were perfect domestic fowl, true to their Dutch origins and tolerant of the wet conditions found in many intensive runs. They enjoyed considerable commercial success as good layers of light brown eggs. As a breed that matures quickly and to a heavy size, males also make good table birds. Their white skin is particularly appealing to European markets. Only one variety of the breed is available, that of the black and white barred cuckoo pattern. The male has slightly lighter colouring than the female bird. The breed is sex-linked, meaning that the sex of its offspring can be determined at one day old by a lighter head spot, even when the North Holland Blue is bred with an unbarred breed. Feather-legged versions of North Holland Blues are now unknown in their homeland and are rare in Britain.

When hybrid hens replaced nearly all commercial breeds and the clean-legged version disappeared from British farms, the exhibition breed almost died out. Dutch North Holland Blues look very like British Marans, so any attempt to admit them to UK poultry standards is still likely to run into the same opposition that the original breeders faced more than 50 years ago.

ESSENTIAL CHARACTERISTICS
Size: Male 3.1–4kg/7–9lbs.
Female 2.7–4kg/6–9lbs.
Varieties: Cuckoo.
Temperament: Very quiet.
Environment: Will tolerate damp conditions.
Egg yield: Early utility strains lay more than 200 eggs per year.

◀ *North Holland Blues have attractive plumage.*

Faverolles

In many ways, the early development of the French Faverolles mirrored that of some strains of the Sussex breed. The breed was developed within a local area to meet the growing demand for quality white-fleshed table fowl in the 19th century. Small-scale farmers would have crossed local indigenous fowl with a variety of other breeds, with each successive cross intended to improve the quality of the breed.

The Faverolles are said to be descended from an indigenous five-toed French breed, crossed with various proportions of Polish, Crèvecoeur and Houdan fowl. The result is a lightweight, very white-fleshed, quality table fowl. The requirement for the breed to have a larger carcass saw the introduction of heavy strains to the breeding pen, such as Dark Dorkings and Light Brahmas. Like their Sussex counter-parts, Faverolles' breeders eliminated any yellow-skinned birds resulting from the Brahma outcross. After several generations, a sizeable table fowl emerged that retained the five toes of its ancestors, along with some vestige of feathered feet from the Brahma. It was the Faverolles' beard

▶ *A classic Salmon Faverolles hen.*

inherited from either its Houdan or Polish outcrossing that became the breed's badge of distinction, to the point when later some saw this to be the breed's most important exhibition feature.

The breed became popular in Britain as a utility breed during the post-World War I expansion in table fowl production. For a while its "cloddy" handling (meaning lumpy – the word is still in the breed's standard) became a feature of both exhibition and utility fowl.

Today the appealing Salmon variety attracts exhibitors as well as those looking for a large and unusual garden fowl. Several other colours exist, in large and miniature versions that fully reflect the breed's ancestry. The bantams are some of the heaviest varieties available.

Selection for beards and fluff may have detracted from the utility merit of some strains, but there have always been breeders willing to select on the basis of early maturity and table type.

Faverolles are strong and hardy birds, with a calm disposition and gentle nature. They can be bullied if kept with more aggressive birds. They are ideally suited to a back garden environment, though they cope well with being confined.

▼ *A large Cuckoo male*

▼ *A pretty Buff bantam.*

ESSENTIAL CHARACTERISTICS
Size: Large male 3.1–4.9kg/7–11lbs.
Large female 3.1–4kg/7–9lbs.
Bantam male 453g/16oz.
Bantam female 396g/14oz.
Varieties: Black, Blue-laced, Buff, Cuckoo, Ermine, Salmon, White.
Temperament: Placid, quiet.
Environment: Copes with confinement.
Egg yield: 150 eggs per year.

Marans

Cuckoo-pattern feathers and very dark brown eggs are synonymous with the breed of fowl standardized as Marans. The irregular and indistinct barring pattern standardized as Cuckoo is typical of breeds where there has been interaction between an Asiatic colour gene and that of the Pencilled varieties of northern Europe. The coloration is typical of many breeds and local populations that evolved in Holland and Belgium, but not in France.

The development of the Marans as a wholly French breed started in the Aubiers region of Deux-Sèvres. By selecting only from those birds that laid the darkest eggs, breeders had, by 1931, made a new breed of hen, the Marans. It is likely that the strains that made up the local breed were crossed with English Game, before later having Asian fowl added to the gene pool.

However, brown eggs imported from France and labelled as Marans, had appeared on the London market from the 1920s, creating doubt about the origin and sequence in importation of parent stock. The breed that was to

> **ESSENTIAL CHARACTERISTICS**
> **Size** (British clean-legged standard): Large male 2.7–3.6kg/ 6–8lbs. Large female 2.7–3.1kg/ 6–7lbs. Bantam male 793–907g/ 28–32oz. Bantam female 680–793g/24–28oz.
> **Varieties:** Black, Cuckoo, Dark-cuckoo, Golden-cuckoo, Silver-cuckoo, Wheaten, White.
> **Temperament:** Docile and calm.
> **Environment:** Hardy bird, tolerates wet conditions.
> **Egg yield:** 150–200 large, chocolate-brown eggs per year.

◀ *A large Cuckoo hen.*

become the most successful brown-egg layer owed much of its British development to Lord Greenaway, a poultry patriot who believed the British poultry breeders should have the resources to compete with the rest of the world. Frustrated by the reluctance of the French to part with breeding stock or even fertile eggs, even after Lord Greenaway met them at the Paris exhibition of 1929, his manager, Parkin, seems to have successfully imported clean-legged Belgian "Coucou de Flandre", some of which laid dark brown eggs. From these he developed his own Cuckoo

Marans. A Marans club was formed in 1950, and the written standard followed soon after. It ensured that poultry keepers had a breed that laid large, dark brown eggs and was also a first-class table fowl. While never really challenging the established laying breeds, strong support from egg exhibitors and specialist egg producers ensured that this version of the Marans became a well-known breed. The exhibitors selected their birds for breeding almost exclusively on the basis of their ability to lay eggs with extremely dark shells. Because of this, some strains may have lost some of their laying productivity.

Marans bantams lay an egg that is just as good as that of their large counterpart, but over the years they have struggled to maintain the required weight. Commercial interests created a Marans-based speckled hybrid that rarely lays an egg as brown as the true breed or in such numbers as a hybrid hen. Meanwhile, the French continued the development of their own original Marans, and standardized the lightly feathered leg and rather different body shape.

▶ *A large Wheaten hen. This is a French standard feather-legged Marans.*

Russian Orloff

The Russian Orloff is an ancient breed; named in honour of a Russian nobleman named Orlov who was responsible for promoting the breed. It originated in Persia (modern-day Iran) and was widely distributed across Europe and Asia by the 17th century; German breeders are responsible for its development. It was included in the American Poultry Standard in the late 19th century, but did not arrive in Europe until the turn of the 20th century. It was introduced to Britain in the 1920s. Today it is considered a rare breed.

It is thought that the Malay breed is partly responsible for the genetics of this bird, although others have suggested that the Thuringian breed has added to the gene pool. In appearance, the Orloff has an owl-like face, with a round head, muffs and a beard. It has a small strawberry comb. This is a large and heavy breed developed as a table fowl; the bantam version was not developed until the 1920s. It is tall, with plenty of feathering, particularly around the head and neck, and generally has a game-like appearance.

The breed has yellow legs, tiny wattles, bay or brown eyes and a small and strong curved beak. The

ESSENTIAL CHARACTERISTICS
Size: Male 3.6kg/8lbs.
Female 3kg/6½lbs.
Varieties: Black, Cuckoo, Mahogany, Spangled, White.
Temperament: Calm.
Environment: Copes with confinement.
Egg yield: 160 light brown eggs per year.

▲ *The Russian Orloff is able to tolerate harsh, cold climates in keeping with its development in the steppes. This is a bantam hen.*

Russian Orloff is a hardy fowl, able to withstand cold temperatures and. The breed has a number of recognized colourways.

Sulmtaler

Like much of the indigenous fowl of Europe, those found in the Styria region of Austria and Slovenia would have been of a lighter weight, primarily egg laying type. While the native Altsteirer has altered little and is rarely found outside of its homeland, the heavier Sulmtaler fowl is now found in ever-increasing numbers over much of Europe.

A 19th-century demand for a heavier fowl saw breeders in the Sulm valley to the south of the city of Graz out-cross to breeds like Cochins, Brahmas, Houdans and Dorkings. The aim was to increase the bird's size and table qualities.

Today it is the handy size bantam version of the resulting Sulmtaler that is rapidly gaining in popularity. While well shown examples have helped promote the breed around the exhibition circuit, their quiet temperament and ability to lay a reasonable number of eggs has seen many find them an ideal small garden fowl.

ESSENTIAL CHARACTERISTICS
Size: Large fowl: Cock 3–4kg, Hen 2.5kg, Cockerel, 2.5kg, Pullet 2kg.
Bantam: Cock 1kg, Hen 0.8kg, Cockerel 0.9kg, Pullet 0.7kg
Varieties: Wheaten and Blue-Wheaten
Temperament: Friendly
Environment: Content in small clean run but happiest when enjoying some garden space
Egg yield: Large fowl 130–150, Bantams 150–190.

▶ *A manageable small crest is an important breed feature, with the natural Wheaten and rather more exotic Blue-Wheaten likely to remain signature varieties of the breed.*

BRITISH MANMADE BREEDS

Dedicated British poultry breeders were renowned for selecting existing and imported breeds and then refining character traits and marketability, to the point where physically they bore little resemblance to the original breeds. Despite this, for a long time there was considerable resistance to the concept of making and naming a new breed of poultry. Sussex poultry breeders, for example, were happy to make use of the local strains of Kent and Surrey fowl as a foundation for the birds that they bred and perfected to become the Sussex breed. William Cook, a skilled poultryman, was the first to standardize and market his newly made Orpington breed in the late 19th century. Exhibitors have since selected and altered his original breed. The Buff Orpington was developed to become a breed in its own right, and is perhaps one of the world's best-known breeds. The Marsh Daisy was created at the beginning of the 20th century, followed a few decades later by the Ixworth. It was not until the 1970s that British breeders created more new breeds. Many of these have been successful locally, but none of those developed have been as commercially successful or as popular as breeds created in other parts of the world.

Orpington

William Cook's introduction of Black Orpingtons in 1886 heralded the first of a family of differently coloured Orpingtons and a new departure in how the British perceived poultry breeds and poultry breeding. Cook, an astute businessman and skilled poultry keeper, having previously advocated and marketed crosses between different breeds, must have seen the acceptance of the new

◀ *This is a heavy breed, with a short back and curvy U-shape, in which the thick feathering almost entirely obscures visibility of the legs. This is a Black pullet.*

> **ESSENTIAL CHARACTERISTICS**
> **Size**: Large male 4.5–6.5kg/10–14lbs.
> Large female 3.4–4.5kg/7½–10lbs.
> Bantam male 907g–1kg/32–36oz.
> Bantam female 793–907g/28–32oz.
> **Varieties**: Black, Blue, Cuckoo, Jubilee, Spangled, White (Buff).
> **Temperament**: Docile, friendly.
> **Environment**: Large space required. Needs protection in wet weather.
> **Egg yield**: 90–150, often quite small, light brown eggs per year.

American breeds as an opportunity to make and market his own breed. Unlike the earlier American breeders, he kept the exact breeding formula of his new breeds to himself, and he saw each new introduction as a new breed rather than a new colour variety. Black Minorca males were crossed with Black Plymouth Rock pullets, and the offspring were mated with a clean-legged Langshan to form the basis of the original Orpington fowl. It is likely that the breed also has some Hamburg ancestry. Cook claimed that Langshans, and his later single-combed Black Orpingtons, descended from urban fowl.

When analyzing how those breeds that were created from a mixture of other breeds have developed, it is usually found that, once the genetic composition is in place, it is how subsequent generations are selected that shapes the future of the breed. Many of the Black Orpingtons were developed to became shorter in the leg, and with enough extra feather and fluff to look like clean-legged Cochins. Once a trend in exhibition poultry has been established it is difficult to stop, and given a tendency for most exhibition strains of heavy, soft-feather breeds to become over-feathery, the Black Orpingtons soon assumed the style and body type that we recognize today. The Buff Orpingtons, which still have their own separate breed club, retained a far more active stance.

It seems likely that the White Orpingtons were derived from sports of the Buff Orpington. Early publications illustrate White Orpingtons as being flat-backed and with rose combs. Interest in this colour strain has varied over time. The Jubilee Orpington was introduced in time for Queen Victoria's Golden Jubilee (1887), but failed to make much progress. Blue Orpingtons are difficult to breed, but every year there are a few Blue large and bantam fowl that are good enough to compete with Blacks.

▼ The Lavender is one of the latest colour varieties, and is so new it has yet to be admitted to the Poultry Club standards.

Cuckoos were never as feathery or fluffy as the Blacks and never really established, but recently, good Spangled varieties have reappeared, to be joined by new Chocolate and Gold-laced varieties.

The Orpington story still has room to develop. They are all very large and heavy birds, with equally large appetites, and require specialized or adapted housing. Few will be even moderate egg-layers, though the breed does go broody and hens make good mothers. Since the breed has been largely bred for exhibition, it is no longer considered to be a table fowl, but there are still those who like to have a large, placid fowl in the garden, and for those with less room, the bantam version, which in some instances is as big as some large fowl breeds, has all the style and character of its larger counterpart.

▼ A White hen.

▼ A Black hen.

▼ A large Lavender hen.

Buff Orpingtons

Committed fanciers may see the extremely heavily feathered and rather rounded shape of the Black Orpington as the ultimate exhibition strain, but for most of the poultry-keeping world, Orpington is synonymous with Buff Orpington. One reason why many with only a passing interest in poultry keeping know the name Buff Orpington is that, unlike the Black variety, it combines an extremely placid disposition with an ability to lay enough eggs to make it worth keeping. It can be found in suburban and country gardens as well as on smallholdings.

William Cook, the breeder, was not keen to publicize just how he bred his Buff Orpingtons, but contemporary reports suggest a Gold-spangled Hamburg was crossed with a Buff Cochin, and the female offspring of that match was mated with a Dark Dorking male. The subsequent off-spring was mated with a Buff Cochin to produce the Buff Orpington. Other reports suggest that he included the non-standardized, but popular, Lincolnshire Buff in his breeding pens. For most of the 20th century, breeders and exhibitors accepted that in order to fulfil the utility aspirations of many poultry keepers, the breed would have to retain a different body type to its Black counterpart.

By the 1980s, few poultry keepers expected pure-bred fowl to lay well, and at the same time, hobby keepers were beginning to look for a bird that would compete with the Black Orpington. British judges officiating at Dutch shows recorded the emergence of strains of Buffs of a type not dissimilar to Black Orpingtons. Importation began from Holland in 1983, and later, from Germany. The result is that the Buff Orpington now being shown is as large and feathery as the Black Orpington. While some may deplore the disappearance of the earlier bird, far more Buff Orpingtons are now being kept. The emerging bantam Buffs are an exact counterpart of the large version. Many bantams weigh more than hybrid hens although they are small. They often lay more, proportionally larger, eggs than their large counterparts, making them the perfect garden companion. Many have fallen in love with the shape, quiet demeanour and colour of this variety of Orpington.

▼ A huge young Buff Orpington male, with a wealth of feather that rivals that seen on the best exhibition Blacks.

▼ A large Buff Orpington hen.

ESSENTIAL CHARACTERISTICS

Size: Large male 3.6–4.5kg/8–10lbs. Large female 3.1–3.6kg/7–8lbs. Bantam male 907g–1kg/32–36oz. Bantam female 793–907g/28–32oz.
Varieties: Buff. The colour may fade in bright sunlight.
Temperament: Docile, friendly.
Environment: Dry run.
Eggs: 120–160 sometimes rather small, tinted to brown, eggs.

Australorp

The Australorp provides one of the best examples of how different strains of a breed can develop to the point where they become separate breeds. When William Cook bred his first Orpingtons, he seems to have selected the original single-combed Black Orpington on the basis of its laying ability. Other breeders would have used varying amounts of Cochin blood in the breed, and it is from these lines that most later Black Orpingtons descend.

All of these strains would have had an early wide distribution throughout Australia, where it is likely that breeders had a different attitude to exhibiting and selection. Selecting fowl from the better-laying families led to many Australian strains of Orpington developing along utility lines. Australian Orpingtons were developed for their utility qualities, and include modest

amounts of Minorca, Rhode Island Red, Leghorn and Langshan in their genetic composition. Some strains of large Australorps are satisfactory egg-layers. An Australian Orpington broke the laying record with 302 eggs in 1920. Few Black Orpingtons would have been even modest egg-layers. First and second generations of importations from Australia in the late 1920s saw all perform well in British laying trials.

The description Australian Orpington was soon shortened to Australorp, and this was eventually the name accepted for the breed by Australian poultry societies.

Later, the breed lost popularity to the best Rhodes and Wyandottes. However, the breed's reputation as being a first-class smallholders' fowl was added to

▲ *A Black bantam male.*

during World War II, when domestic poultry keepers found them to be excellent layers even when fed on wartime rations. Competition ensures that all exhibition breeds provide individual challenges, but the bantam version now found in Black, Blue and White varieties has gained a reputation as being among the best miniature fowl for those with limited space. In common with the strains of most breeds now being kept by enthusiasts, few lay as many eggs as their ancestors, but with some selection this quiet fowl could still be a useful addition to many gardens.

▶ *A large Australorp hen with the firm stiff tail, rising from a neat cushion, that distinguishes this variety from the fluffier Orpingtons.*

ESSENTIAL CHARACTERISTICS

Size: Large male 3.8–4.5kg/ 8½–10lbs. Large female 2.9–4kg/ 6½–9lbs. Bantam male 1kg/36oz. Bantam female 793g/28oz.
Varieties: Black, Blue, White.
Temperament: Calm, docile.
Environment: From free range to intensely housed.
Egg yield: 120–180 tinted to mid brown eggs per year.

Lincolnshire Buff

Poultry from Lincolnshire, England, was used extensively to supply meat to the London food markets. It is likely that Dorking-Cochin crosses formed the basis of many breeding pens in that county, but the breed remained unrecognized outside its home country.

▶ *The five toes and body shape of this adult male reflect the Dorking part of the breed's ancestry.*

In the 1980s, Riseholme College, an agricultural college in Lincolnshire, began a program of refining the breed. This work was continued by local fanciers until such time as the breed could be bred to an agreed standard. By the time a breed club had been formed in 1995 the breed would have found new friends among those for whom the Buff Orpington had become far too feathery to meet their requirements for an all-round utility fowl.

The Lincolnshire Buff may remain largely confined to its home county, but seems to be here to stay, albeit in large form only. In appearance, the breed has a long back and carries its tail at a low angle. It has five toes, and tighter feathering than the Buff Orpington, whose previous role as a an all-round useful and attractive, buff-coloured fowl it is beginning to replace. The two-toned buff pattern is one of the least demanding to breed and maintain.

ESSENTIAL CHARACTERISTICS
Size: Large male 4–4.9kg/9–11lbs.
Large female 3.1–4kg/7–9lbs.
Bantam male 1.1kg/40oz.
Bantam female 963g/34oz.
Varieties: One only.
Temperament: Quiet.
Environment: Free range.
Egg yield: 120–150 medium tinted eggs per year.

Norfolk Grey

The Norfolk Grey was the creation of Fred Myhill of Norwich, whose poultry-breeding enterprises were interrupted by a spell in the World War I trenches. Norfolk Greys were widely advertised in the poultry press but the exact origins of the breed are unclear. Leghorn and Game, with an infusion of either or both Black Orpington or Australorp, seem to have been the genetic base of stock

ESSENTIAL CHARACTERISTICS
Size: Male 3.2kg/7lbs.
Female 2.7kg/6lbs.
Varieties: One only.
Temperament: Quiet.
Environment: Free range.
Eggs: 140–180 tinted to brown eggs per year.

being offered in the 1930s. The breed never became popular, and little seems to have been seen or heard of the breed until the formation of the Rare Breed Society in 1970, when four birds not unlike large Grey Old English Game appeared at shows. These four birds were responsible for the breeding pens that are known today. While still rare, this regional poultry breed has made something of a recovery.

The breed is an excellent meat bird, no doubt the game element of its genetic make-up accounts for its good flavour. The Leghorn ancestry ensures that the breed is a reasonably productive egg-layer.

Norfolk Greys are black with silver hackles, black or grey legs, a single

◀ *A Norfolk Grey male looking very like a large, table-type, Old English Game bird.*

red comb, lobes and wattles. The bird is classed as a heavy breed, although its weight suggests otherwise. This is a hardy breed that likes to forage. Like other British breeds, the Norfolk Grey was marketed for a new audience between the two world wars that wanted a smaller plump carcass.

Marsh Daisy

The Marsh Daisy evolved, like several other minor British breeds, out of the search for an ideal dual-purpose fowl, which saw poultry keepers introduce both Indian and English Game breeds to their breeding pens. John Wright of Marshside, Lancashire, originally used both Old English Game and Malays, along with Hamburg and Leghorns, and then bred their offspring together as a closed flock, to ensure there were no further introductions of new blood into the flock. Breeding from older hens probably resulted in the breed's later recorded longevity. In 1913, having previously turned down offers for his stock, he sold some hens to Charles Wright from Doncaster, who first introduced a pit game cock and later added Sicilian Buttercup to the mixture. From this beginning, the Marsh Daisy breed developed into one which has gained a surprising degree of commercial success. A reputation for laying large eggs into old age could be the result of the original breeder only breeding from old hens. This, combined with a plump game conformation, saw the breed remain on some farms until just before World War II. Now rare, this green-legged breed is still seen at the occasional show.

▲ *The Marsh Daisy is considered to be a rare breed.*

ESSENTIAL CHARACTERISTICS
Size: Male 2.4–2.7kg/5½–6lbs.
Female 2–2.4kg/4½–5½lb.
Varieties: Buff, Red-wheaten.
Temperament: Active.
Environment: Garden.
Egg yield: Once good, and recorded in 1939 as laying large eggs into old age.

Ixworth

The Ixworth was created by well-known waterfowl and duck breeder Reginald Appleyard in Ixworth, Suffolk, in the UK, in order to fulfil the pre-war commercial demand for white-fleshed poultry, using the carcass of the Indian or Cornish breed. Starting with Jubilee Indian Game and White Sussex, Appleyard is said to have added both Pile Old English Game and White Orpington blood to his breeding pen.

Not wishing to be influenced by appearances, he is said to have made all his final breeding selections in the dark, by feeling the muscle, bone and flesh structure of each bird. Not standardized until 1939, few Ixworth birds survived World War II. Those that are exhibited today are often large and excellent examples of the Ixworth. The breed is also an excellent egg-layer.

◄ *An Ixworth male displays much of the useful conformation of a game fowl without the excessive bone of some Indian Game birds.*

ESSENTIAL CHARACTERISTICS
Size: Male 3.8–4kg/8½–9lbs.
Female 2.7–3.1kg/6–7lbs.
Varieties: White.
Temperament: Quiet but game like, hardy.
Environment: Free range.
Egg yield: 150–180 eggs per year.

Only one colourway, White, is available in the Ixworth breed. The birds have pink legs and a pink beak. The eyes are orange and it has a red pea comb.

The breed is an active forager and prefers a free-range environment. The Ixworth is a hardy breed that is reported to provide the best-quality meat of any modern pure breed.

LONG-TAILED JAPANESE BREEDS

Native Japanese poultry known as "jitory", have a body type similar to the native fowl of northern Europe and, in some breeds, to the early feathery forms of English game fowl. As poultry meat and eggs were not widely consumed in Japan before the mid-19th century nearly all early poultry selection in that country was made purely on the basis of the perceived aesthetic qualities of the bird. The native poultry had single combs and white lobes. They were crossbred with breeds imported from China and Thailand to produce a group of fowl with long tails.

Japanese poultry keepers selected their birds for their long tail feathers. Ten long-tailed Japanese breeds with varying combinations of comb forms and either red or white lobes are known. Once a non-moulting gene within the Japanese group of fowl had been identified, poultry could be bred with tails that could reach quite extraordinary lengths. The incorporation of the non-moulting gene into the genes of the native Onaga-dori long-tailed breed meant that that breed could only survive in the most artificial of environments.

Yokohama/Phoenix

Long-tailed Japanese fowl arrived in Europe in 1860, and from the Onaga-dori breed and the Minohiki breed, German poultry breeders created the Yokohama breed. Early illustrations of the new breed show a diversity of comb and type that could suggest it has a genetic relationship with several Japanese long-tailed breeds.

In Britain new poultry breeds were generally named after their port of dispatch: in this case, the breed was sent from Yokohama in Japan. In Germany the single-combed versions became known as Phoenix fowl, after a mythological

▶ *A large Silver-duckwing male has the non-moulting mutation that allows some birds to attain very long tails. The breed requires attention, since the tail may hamper natural movement.*

bird said to rise up from a fire's ashes. Later German breeders called their long-tailed, walnut-combed, game-type fowl Red-saddled Yokohama. That breed had features in common with, but had no exact counterpart, among the long-tailed breeds now being kept in Japan. The brilliant

◀ *A single-combed Yokohama exhibiting the breed's characteristic wealth of saddle and hackle feathers.*

red-backed males and uniquely pink chequered-breasted females make Red-saddled Yokohamas one of the most attractive of the long-tailed group.

YOKOHAMA ESSENTIAL CHARACTERISTICS

Size: Large male 1.8–2.3kg/4–5lbs.
Large female 1.1–1.8kg/2½–4lbs.
Bantam male 566–680g/20–24oz.
Bantam female 453–566g/16–20oz.
Varieties: Game colours.
Temperament: Tame if handled.
Environment: Very long-tailed males will need total protection.
Egg yield: 80–100 small white eggs.

ONAGA-DORI

Onaga means long tail and *dori* translates as fowl. The development of these long tails seems to have been encouraged by the use of long tail feathers in the processional train for the attendants of the Shogun's court in Japan. In total, there are up to 25 pairs of long tail feathers in a complete Onaga-dori tail. These are enhanced by up to 200 non-moulting saddle hackle feathers. Records from 1830–43 show the breed had a tail that reached 3m/3¼yds long. Between 1863 and 1922 an extra metre was added to the length of the tail. By 1930, several birds were reported with tails greater than 6m/6½yds long, and by the late 1950s it had increased to 8m/8¾yds.

The male Onaga-dori's long tail develops over many years; and then only when the bird is kept in such a way as to let the tail continually hang down. Since the breed moults if the bird is allowed to spend much time in more usual poultry conditions, it is unlikely that the western world will ever see this sort of development in their Onaga-dori breeds.

▶ *Some examples of the legendary Onaga-Dori of Japan have tails that are several metres long. Such lengths are attained by older male birds, many of whom will spend most of their life standing on a high perch, with the tail allowed to hang down freely.*

By the early 1900s, the single-combed breed standardized in Britain as Yokohama was popular enough to have its own breed club. Most of these birds had white lobes and would have been very similar to the Japanese Shokoku breed. Some illustrations show red-faced, pea-combed breeds that suggest the importation of other Japanese varieties, but as the similarly shaped Sumatra Game had also found its way to Europe, some of the early long-tailed examples could be the result of early outcrosses.

An importation into Germany in 1976 probably represents one of the few times the Onaga-dori has been seen in Europe. While these produced offspring with very long tails, the expertise to delay moulting by more than a couple of years seems to have eluded exhibitors. Animal welfare considerations are likely to see those males kept in more natural conditions in Europe with a reasonable amount of trailing feathers. In fact, while they are likely to be seen primarily as

RED-SADDLED YOKOHAMA
ESSENTIAL CHARACTERISTICS

Size: Large male 1.8–2.3kg/4–5lbs.
Large female 1.1–1.8kg/2½–4lbs.
Bantam male 566–680g/20–24oz.
Bantam female 453–566g/16–20oz
Varieties: Game colours.
Temperament: Active.
Environment: At its best on free-range grass. Some strains are intolerant of confinement.
Egg yield: 80–100 small white eggs.

exhibition fowl, all forms of the large Japanese long-tailed fowl standardized in Europe (including the bantams made from them) have more

range and freedom than most breeds. The bantam forms of both the Red-saddled and single-combed breed seen at recent shows seem to have been bred from smaller examples of large versions, and while perhaps on the large side, have displayed an exceptional wealth of feather.

◀ *A Red-saddled Yokohama bantam male. Many examples shown as bantams exceed the standard weights. This breed is unknown in Japan.*

Ohiki

Japanese long-tailed poultry breeds are found across a considerable size range. The concept of a miniature or bantam version bred to conform to a percentage of the size of a large version does not really exist in Japan, where each breed is seen as having its own ideal weight. The long-tailed Ohiki, however, is a true bantam, one that has much in common with the Japanese Bantam/Shabo breed. It is thought to be developed from the Onaga-dori breed crossed with the Japanese bantam. Like the latter breed, the Ohiki has very short legs and a long, trailing saddle hackle, with long, soft downy feathers beneath, meaning that the exhibition Ohiki has to spend much of its life confined to a clean, shaving-covered pen in order to protect the tail. The tail feathers can measure any length from 60–150cm/2–5ft and are held initially upright from the body, then dragging behind the bird at a low angle. The breed has an abundance of plumage, including hackle feathers, giving it the appearance of a full and round breed. The wings are long and carried low to the ground. The bird walks with a horizontal carriage.

◀ *The breed has a single comb, white earlobes and willow legs. This is a Black-red male.*

ESSENTIAL CHARACTERISTICS
Size: Male 0.9kg/2lbs.
Female 0.7kg/1½lbs.
Varieties: Black-breasted Red, Ginger, Gold-duckwing, Silver-duckwing, White.
Temperament: Calm.
Environment: Tolerant of confinement, males will need a dry run with a clean floor.
Egg yield: Kept for ornament.

Kuro Gashiwa

Another long-tailed breed from Northern Japan, with feathering that drags behind the bird. The bird is completely black, with slate-coloured legs and feet. The most distinguishing feature of this bird is its crowing, which is long and deep. Long-crowing roosters have over the years had almost cult status among enthusiasts in many countries, where length and tone of crow is deemed to be at least as important as breed type. Standardized forms are found in Germany, Belgium and Brazil, where they have an Asiatic game form. All have a sloping back like the Japanese Kuro Gashiwa. Japanese legend has it that such poultry was first introduced from China to act as alarm clocks – a long, loud crow would have been seen as a desirable attribute.

▼ *The Kuro Gashiwa is an unusual addition to the garden.*

ESSENTIAL CHARACTERISTICS
Size: Male 2.7kg/6lbs.
Female 2.3kg/5lbs.
Varieties: Black.
Temperament: Friendly.
Environment: Free range, but neighbours may complain about noisy crowing.
Egg yield: 50–100 white or tinted.

Sumatra

Named for the Indonesian Island from where it originates, this poultry breed has had a chequered development since arriving in America in the mid-19th century. It was imported to Europe from Canada in the late 19th century. This is a game bird used for cockfighting in its native country, though it is better suited to crossing with game birds for the genetic heritage it provides its offspring. The breed's arrival in Europe coincided with the end of cockfighting, and thus it never gained the popularity of other game birds. The triple spurs found on some males may be seen as an indication of purity but are not an essential part of the breed standard.

However, at the time of its importation there was interest in long-tailed Japanese breeds. Some breeders saw the tail of the Sumatra as the breed's most important feature. Its tail is held low, and in the male has a very impressive long sweep made up of many pointed feathers. These long, pheasant-like tail feathers may have assisted the Sumatra's semi-domesticated ancestors in flying out of harm's way. This breed can still fly.

Sumatras have an upright, pheasant-like carriage, small pea comb, almost non-existent wattles and small earlobes. It is their beetle-green feathering that sets this breed apart from any other. A dark or plum-coloured face is a breed characteristic.

Few breeds will enjoy total freedom to range or sleep in trees as they may have done in their native environment. Given a fox-free run Sumatra are capable of looking after themselves. The Black Sumatra has no equal as a broody, and is often crossed with the Silkie breed to make the perfect foster mother for exotic chicks. With a bantam version now available, those with less room will be able to keep one of the poultry world's most interesting, and slightly unusual, fowl.

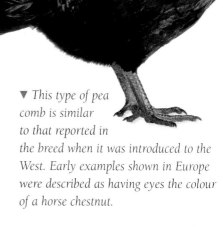

▼ *A large black hen with firm game-like outline and almost whipped tail.*

▼ *This type of pea comb is similar to that reported in the breed when it was introduced to the West. Early examples shown in Europe were described as having eyes the colour of a horse chestnut.*

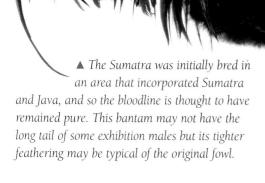

▲ *The Sumatra was initially bred in an area that incorporated Sumatra and Java, and so the bloodline is thought to have remained pure. This bantam may not have the long tail of some exhibition males but its tighter feathering may be typical of the original fowl.*

ESSENTIAL CHARACTERISTICS

Size: Large male 2.2–2.7kg/5–6lbs. Large female 1.8–2.2kg/4–5lbs. Bantam male 737g/26oz. Bantam female 623g/22oz.

Varieties: Black, Blue, White.

Temperament: Bold, but at times, wary.

Environment: Exhibition males may want some protection. Adult breeding stock will enjoy free range.

Egg yield: 100–130 smallish white or tinted eggs per year.

MODIFIED FEATHER BREEDS

Feathers with changes in their structure are likely to occur from time to time within any group of fowl. Some of the more extreme examples are associated with early civilizations. The long pheasant-like tail feathers of the Sumatra breed may have helped its semi-domesticated ancestors to fly out of harm's way, but the extremely long tail feathers of the Onaga-dori breed, which are enhanced by a non-moulting gene, mean that the breed would only survive in the most artificial environment. Similarly frizzle-feathered fowl are a prime example of a modification that would have little chance of survival other than in the most

protected environment. Silky-feathered fowl would not have retained their unique characteristics without selection, and the unique genetic package that includes black skin and dark flesh that has been selected and standardized as the Silkie breed would have soon become diluted by crossing it with fowl of different-coloured skin. At the other end of the scale, Naked Neck sports seem to have occurred regularly and make up a large proportion of the fowl that survive in quite harsh environments. This group contains some of the most genetically interesting and spectacular varieties.

Silkie

The origin of the Silkie is wholly Asiatic; Marco Polo, the Italian traveller and merchant who introduced Asian cultures to Europe, first sighted fowl with similar feathers to the Silkie breed towards the end of the 13th century, while travelling through Asia. He reported them as "Hens which have hair like cats, are

black and lay the best of eggs". As did later 16th-century reports, mainly from Sumatra and Java, he noted that the fowl had black skin and legs. These features are still retained in the breed standard, along with its

> **ESSENTIAL CHARACTERISTICS**
> **Size**: Large male 1.8kg/4lbs.
> Large female 1.4kg/3lbs.
> Bantam male 623g/22oz.
> Bantam female 510g/18oz.
> **Varieties**: Bearded and non-bearded.
> Black, Blue, Buff, Partridge, White.
> **Temperament**: Docile and friendly;
> the perfect pet, but often persistent
> broodies.
> **Environment**: Needs a covered run.
> Wet, muddy and hot conditions are
> unsuitable. Copes with confinement.
> **Egg yield**: 100–120 cream eggs
> per year.

▼ *A Black unbearded hen.*

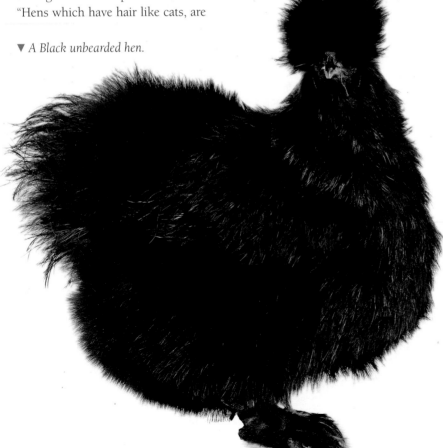

feather silkiness. Since the breed is small it does not have the carcass of a meat bird. Its black skin, flesh and bones are not to Western tastes, but in China the meat is considered to have all manner of health-enhancing and medicinal properties.

Silkies, like their name suggests, are associated with silky feathers that feel like down. It is a feature that was once unique to this breed. In the best specimens, such downy feathers almost replace standard feathers. This attribute, however, ensures that Silkies cannot fly. Its other distinctive characteristics include a modified walnut-type comb, and five feathered toes. The silky feather type has now migrated to other breeds that do not have these additional characteristics. An unusual feature of this breed is the turquoise earlobes, which are quite striking.

Unlike Frizzles, when mated, two good silky feathered specimens will always produce all their offspring with this same feather characteristic. While the breed does not lay a massive quantity of eggs, broodiness is a strong characteristic. In fact, the smaller versions are a broody of choice for those looking to naturally incubate eggs from other poultry, or for small waterfowl breeds that do not go broody. Silkies naturally make good mothers. Some breeders use pure-breed Silkies to produce crossbreeds that almost always inherit the legendary maternal instincts.

Silkies have been a feature of our farmyards and gamekeepers' rearing pens for a long time, but they are not widely represented across broad geographic regions. Often these slightly undersized birds are impressive at first sight because they appear to be so unusual.

Without a bantam counterpart, and being a small breed, Silkies were often considered to be large bantams. The creation of a bantam version (with weights as low as 510g/18oz for females), which is rare in Europe but popular in the USA, helps to dispel this misconception. Dedicated work by skilled breeders may make bigger strains of the large breed in the future (with weights up to or exceeding 1.8kg/4lbs for males). Japanese breeders have spent years perfecting their bantams so that they have become very silky.

White birds generally dominate the exhibition classes. However, the original black varieties can be just as

◀ *A White male Silkie. It is not just the silky or almost wool-like feathers that have long defined the Silkie as a breed. The complete package includes black skin, five toes and a mulberry or walnut comb.*

stunning. The arrival via Europe of a Bearded variety that has been perfected in America will add future interest and variety to the show scene. This new variety is a "cloddy" type, which lacks the bright-eyed and alert stance of the traditional type. Not everyone will want to exhibit Silkies, however, and this ornamental breed makes a perfect pet. Docile and calm by nature, the breed should be put with non-aggressive birds since it is susceptible to bullying.

▼ *A muffed and bearded black Silkie bantam hen.*

Frizzle

The Frizzle is standardized as a feather type which may be bred into any poultry breed. In Britain, the Frizzle is also classified as a distinct breed. Frizzled feathers curve backward, whereas standard feathers curve forward. They provide the bird with a shaggy, rather than smooth, appearance. The perfect, wide, reflexed feather remains a challenge for breeders to perfect. This feather modification provides a bird with little chance of survival in any other

▲ *A frizzled Pekin male.*

▲ *A large White Frizzle.*

ESSENTIAL CHARACTERISTICS

Size: Large male 3.1–3.6kg/7–8lbs. Large female 2.2–2.7kg/5–6lbs. Bantam male 680–793g/24–28oz. Bantam female 566–680g/20–24oz.
Varieties: Many.
Temperament: Quiet, friendly.
Environment: Good housing and a dry run.
Egg yield: 80–150 white tinted or brown eggs per year.

than the most protected environment. Frizzled fowl were described as early as 1600. Early naturalists agreed that frizzled feathers were an Asiatic characteristic, possibly originating in Japan. It is thought that frizzled breeds came to Europe via the Dutch trading port of Bantam. Nearly all imported birds with this feather characteristic conformed to a type later categorized as true bantam.

The frizzle feather can often be found in commercial strains of large heavy breeds like Rhode Island Reds and White Rocks, as well as broiler breeding strains. These large Frizzles, in all other respects, match the standard for their particular breed. All large frizzled fowl are classified as heavy for exhibition purposes. The original bantams are now shown as miniatures of the heavy versions, rather than true bantams.

As poultry keeping becomes an increasingly international hobby,

the world is likely to see more breeds adopt a frizzled variety. It is now usual to see frizzled Polish, and perhaps more controversially, frizzled Cochins and Pekins, and long established breeders see this as being an unwelcome departure from their standard of perfection. Many poultry fanciers will get pleasure from a quaint and unusual variation of the standard breed.

FRIZZLED FEATHERS

The frizzle feather type provides an excellent example of Mendel's theory, with regard to dominant and recessive genes. Two mated Frizzles will produce offspring in which 50 per cent have frizzled feathers that match the parents. Of the rest, 25 per cent will have a standard feather type, and the balance will have feathers that are over-frizzled, meaning that they will be so extremely curled and fluffy that the chicks would be unlikely to survive in a challenging environment. This over-frizzled category, however, if crossed with their standard feathered siblings, can be used in a breeding program that allows the frizzled feather gene to "migrate" to standard-feathered breeds.

▶ *A Buff Frizzle male bantam.*

Transylvanian Naked Necks/Turken

The most dramatic changes in the feather structure of poultry are most likely to be perpetuated with human assistance. However, the opposite is true in the case of the Naked Neck breeds. Related groups of these birds seem to occur spontaneously. The most significant visible characteristic of the Naked Neck, as its name suggests, is that the neck area is devoid of feathering. These breeds also have less feathering across the breast bone. Most of the diverse populations of Naked Neck fowl seem to have emerged in response to difficult conditions. Such conditions do not add to a bird's chance of survival, and in many cases the naked neck characteristic will prove counterproductive to an ideal life in challenging, or even habitual, environments. Naked Neck birds are more likely to survive in domestication.

In spite of the prefix Transylvanian given to the breed name, populations of fowl with nearly or completely naked necks have a wide distribution, though there may once have been some connection between the breed and Central Europe. It may have originated in Hungary, though it was largely developed in Germany. Its alternate name of Turken is thought to have arisen from a belief that this breed, with its naked neck, was a hybrid between a chicken breed and a turkey.

In the early 20th century, an agricultural adviser to Grenada reported that the breed was preferred in the West Indies to imported birds because of its superior hardiness and other qualities. It seems this hardiness also attracted subsistence farmers in the French-African colonies, who kept Naked Necks under the name Barbary Fowl. Later, French scientists noticed that as table fowl this breed was practically devoid of visible surface or skin fat.

The fat issue was of great interest to health-conscious consumers, and led French poultry geneticists to breed commercial strains of the breed. The Naked Neck breed has since been found to be linked with a useful gene that disperses fat through the dry meat areas in table birds. It also gives a wonderful texture to the meat.

◄ *The breed is said to have survived the Wartime siege of Malta, because their vulture-like necks enabled them to dig into rubbish for food.*

▲ *The neck of the Transylvanian Naked Neck breed should be a deep red colour, in order to achieve high marks at exhibition. The skin turns red when exposed to sunlight.*

Hybrid table strains that retained the characteristic naked neck could be used to create new strains that, with selection, may conform to the exhibition standard for Naked Necks. This demonstrates the value of conserving the diversity contained in the most unusual forms of poultry. This breed is known to be resistant to many poultry diseases. Naked Necks have fewer feathers than other breeds, including those over the rest of the body.

The bantam versions make both unusual exhibits and curious additions to any poultry yard. This is a good dual-purpose bird converting feed to eggs at an efficient rate.

ESSENTIAL CHARACTERISTICS
Size: Male 3.1–3.6kg/7–8lbs.
Female 2.4–2.9kg/5½–6½lbs.
Varieties: Black, Buff, Red, White.
Temperament: Friendly.
Environment: Free range or confinement. Needs protection in cold weather.
Egg yield: A good layer of light brown medium-size eggs.

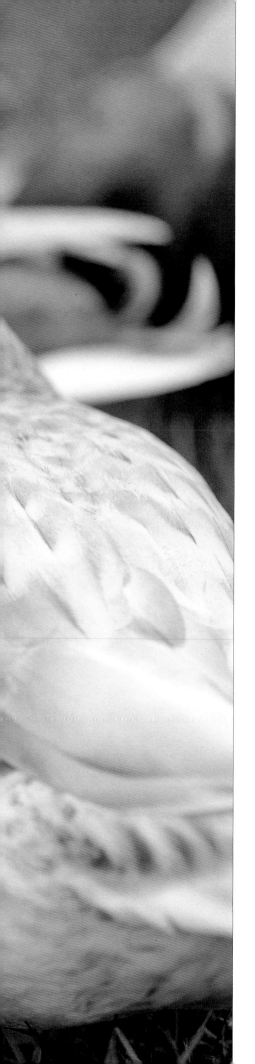

A DIRECTORY OF OTHER DOMESTIC FOWL

Ducks, geese and turkeys are bred to be shown at exhibitions by specialist and hobby enthusiasts. They are also kept as domestic pets. For small-scale and amateur breeders, the appeal of keeping these birds may stem from a fascination with the character traits, personality and physical appearance of the species. It may represent the chance to own an old-fashioned breed familiar to one's grandparents, as well as the opportunity to keep the breed alive for future generations. Ducks, geese and turkeys are all quite vocal species, which is worth bearing in mind if you intend keeping them in an area close to neighbours. Each breeder will look for specific features that he or she considers important in their fowl, then select the best of their stock. Selected birds will be used to breed the next generation in an attempt to produce offspring that perfectly fit the accepted breed standard for exhibition fowl.

▲ *Ducks make unusual pets, but are most usually associated with a rural, rather than urban, environment.*

◄ *Call ducks make charming pets and can be accommodated in a relatively small space, such as a large garden.*

DUCKS, GEESE AND TURKEYS

All of today's breeds of domestic fowl would have been identifiable in varying shapes and sizes by the middle of the 19th century. It was at this time that written exhibition standards for domestic waterfowl were drawn up, and breeders perfected their birds.

There are less breeds of turkeys, ducks and geese than there are of chickens, but this has not quashed breeder interest. Standards of perfection are high among this diverse group of birds.

Ducks

All the domestic breeds of duck, with the exception of the Muscovy (*Cairina moschata*), descend from the wild mallard (*Anas platyrhynchos*). The mallard is a migratory bird and breeds with all known duck species, including domesticated ducks.

It is not known for certain when ducks were first domesticated. It seems likely that as soon as humans started growing crops and living in agricultural settlements, wild mallard ducks arrived to share the harvest. Their presence was tolerated because the occasional bird provided meat, feathers for padding and insulation, and eggs for eating. Ducks would eventually have laid second and

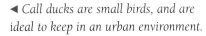

third clutches of eggs in order to compensate for the loss of eggs. Over time, more productive ducks would have evolved and became the basis of lightweight laying breeds.

Evidence suggests the ancient Egyptians may have bred ducks for food and also to sacrifice in religious ceremonies. In South-east Asia they were being reared before 500BC, and roast duck was certainly a dish prized by the Romans. Centuries later, the Normans introduced selected heavy

▼ *Rouen ducks have the same colouring as the wild mallard.*

◄ Call ducks are small birds, and are ideal to keep in an urban environment.

strains of mallard-coloured and white domestic ducks for the table. Monastic records from this time indicate that waterfowl were more important to the diet than poultry. While many of the ducks that were bred had a modified mallard pattern, birds with white feathers and bill were understood to represent a bird with quality meat.

Before 1890, flocks of ducks were hardly ever kept as egg-layers, even though domestic ducks were more prolific egg-layers than their wild ancestors. The tall, ultra-lightweight "Runner"-type ducks first introduced by the Dutch from the East Indies allowed the creation of breeds that laid greater numbers of eggs than did hens. Within 30 years, vast numbers of lightweight ducks, all related to, or partly descended from, Indian Runners were kept. Food shortages during and after World War I brought attention to lightweight ducks that could lay far more eggs than any hen of the period. The Khaki Campbell and Indian Runner became the breeds of choice for small-scale poultry keepers. Yellow-billed Pekin ducks imported from the East offered the possibility of strains that both laid well and grew quickly, qualities desirable to breeders.

Geese

Traditionally associated with the storybook farmyard, geese are large and strong birds, renowned for their noisy and aggressive behaviour. They are kept for their meat and eggs.

In Europe, North Africa and western Asia, domestic geese are descended from the wild Greylag goose (*Anser anser*). Those of eastern Asia are thought to be descended from the wild Asiatic Swan goose (*Anser cygnoides*). Known as Chinese geese, they can be distinguished from European geese by the large knob at the base of the bill. However, both kinds have been widely introduced in modern times, and many flocks in both regions now contain either type.

The recent boom in small-scale poultry keeping has sparked similar heightened interest in keeping geese. Specialist breeding companies are able to supply goose eggs to hatch, or goslings just a few days old, to poultry keepers who want to keep geese for meat and eggs. European breeds are particularly fashionable. For pure breeds the best source of stock is a small-scale breeder.

Turkeys

The wild turkey, *Meleagris gallopavo*, originated from Mexico. There it was domesticated by the Aztecs and other

▼ *Pilgrim geese have a useful auto-sexing characteristic.*

Mesoamerican peoples for meat and eggs, and also for its feathers, which were used for decorative purposes. The species was brought to the attention of Europeans through the Spanish colonization of the Americas in the early 16th century. From its homeland, the turkey was introduced to Spain, and from there, to the rest of Europe. All domestic turkeys are descended from the wild species brought to Europe. Following its introduction into Europe, a number of distinct breeds were developed.

Domesticated turkeys were taken back to the USA by European

▲ *Beltsville White turkeys are an American breed and now quite rare.*

migrants almost two centuries later. These immigrant settlers bred and selected heavier domesticated strains of turkey that became associated with the feast of Thanksgiving, and later with Christmas dinner. The breed is now popular with commercial livestock farmers the world over because it is cheap to rear for the amount of meat it yields.

▼ *Sebastopol geese have softly curled feathers on their bodies.*

DUCKS

From the tiniest Call to the large and feisty Rouen, ducks are an attractive species to behold, and will undoubtedly add substance and character to any backyard. With such an array of sizes, temperaments and physical appearances, there is sure to be a duck variety that will suit your purposes, whether you want to keep ducks for meat, eggs, or ornamental value. All ducks need space to wander and a body of water in which to swim, although small ducks need no more than a paddling pool-size area in which to dip their feathers. For those with a small garden and no expanse of water, the small ducks may appeal.

In exhibition terms, ducks are classified according to their weight – heavy, medium and light. However, accepted bird weights differ from country to country. As with poultry, new breeds have been created to fulfil a commercial requirement for meat or eggs, and the utility varieties frequently have slightly different characteristics from the exhibition types.

By definition, ducks have a broad and low body shape, usually a long neck, strong scaly legs and bill, and wings that are set back on the body, supported by strong muscles. Males and females have a different appearance.

Aylesbury

Like all ducks except for the Muscovy, the Aylesbury is descended from the wild mallard. A relatively recent introduction to the duck family, the Aylesbury was developed at the turn of the 19th century for its utility qualities, and later for the exhibition scene. The true Aylesbury has snow-white plumage with a pinky-white or flesh-coloured bill and legs, the latter placed mid-way along its undercarriage, giving it a boat-shaped body and horizontal carriage. The keel should almost touch the ground when it stands upright.

By the beginning of the 20th century, Aylesbury ducks had become one of the world's best-known breeds. Nowadays, there are two distinct types of Aylesbury: exhibition birds, which remain true to the written standard for the breed, and those developed for the table. Often the latter are Aylesbury-Pekin crosses, and are distinguished by their bright yellow beaks. To create these crosses, some breeders in the Aylesbury district were recorded as having introduced a percentage of

Pekin blood into their strains. The Aylesbury-Pekin cross became the duck of choice for those seeking to supply the mass market, and after generations of selective breeding, this is the basic duck now used to produce most of the world's duck meat. However, as with other breeds developed for the quality of their meat, white-fleshed skin is more appealing to the consumer, and white feather stubs produce the least amount of blemishing to the skin.

Utility breeds of the true Aylesbury are desirable for those who want duck eggs and meat, although most birds are kept as garden pets. Utility lines may produce 150 eggs per year, but most produce considerably fewer. Most convert food to meat at a good rate and are ready for the table when they reach 1.8–2.2kg/4–5lbs, usually at eight weeks.

Since the Aylesbury is one of the largest duck breeds, it needs considerable space. Its size prohibits it from mating on land, and so this breed needs access to deeper water for mating. Aylesbury fertility can be a problem, but true enthusiasts think this a premium worth paying.

ESSENTIAL CHARACTERISTICS
Size: Drake 4.5–5.4kg/10–12lbs. Duck 4–4.9kg/9–11lbs.
Varieties: One (white).
Temperament: Slow, ponderous.
Environment: Needs space, and water for mating.
Egg yield: 150 blue, green or white eggs per year. Exhibition strains lay fewer.

◄ *Large table-type Aylesbury exhibition strains hold their breast bone, or keel, almost horizontal, and nearly touching the ground.*

Pekin

The Pekin duck, which had been bred in China for centuries, was introduced into the United States in 1873. Often referred to as the Long Island duck, it became widely popular there. Soon British breeders were importing these birds directly from China. The Pekin is only slightly smaller than the Aylesbury, is a better layer and is more fertile. However, although it is well-fleshed and matures quickly, largely due to a prejudice against yellow skin and other pigmentation it has always been considered to be inferior to the Aylesbury in terms of meat.

The Aylesbury-Pekin cross became the breed favoured by commercial breeders, who supply the market for table fowl. It remains the strain used to produce most of the world's duck meat, and Chinese Pekin duck remains a popular speciality dish.

The Pekin has white plumage with a hint of yellow running through it. With a bright orange bill and legs, it is an attractive bird. The legs are placed well back, giving the bird an upright stance with a characteristic turned-up rump. Some exhibition examples of Pekins have an almost primrose hue. Like exhibition Aylesburys, commercial birds have more flesh and the birds can reach up to 5kg/11lbs.

Commercial Pekins make prolific egg layers, though broodiness has been largely bred out of the stock. Often the ducks may abandon their eggs before they hatch, so hatching eggs in an incubator is a preferable method of ensuring the next generation survives.

ESSENTIAL CHARACTERISTICS
Size: Drake 4kg/9lbs.
Duck 3.6kg/8lbs.
Varieties: White, with exhibition strains having a primrose tint.
Temperament: Placid.
Environment: Needs plenty of space and a large house.
Egg yield:150–200 white eggs. Exhibition strains lay fewer white, blue or green eggs.

◀ *Exhibition Pekin ducks have a shape that appears far heavier than they actually are.*

Both exhibition Pekins and Aylesburys are less well-fleshed than their commercial ancestors. Yet they remain a useful genetic resource as the commercial gene pool becomes ever more interbred.

Cayuga

Coveted for its beautiful iridescent black plumage, which is tinted beetle-green, the Cayuga duck is an ornamental breed that is also kept for its meat and eggs. In appearance it has a black bill, shanks and toes.

ESSENTIAL CHARACTERISTICS
Size: Drake 3.6kg/8lbs.
Duck 3.1kg/7lbs.
Varieties: Black.
Temperament: Quiet but active.
Environment: Happiest where it can have access to a large pond.
Egg yield: 100–150 blue, green or white eggs per year.

Adults typically weigh up to 3.6kg/8lbs. Originating in New York and considered rare, this breed is increasing again in popularity.

It is a docile and hardy breed that needs plenty of space. Cayugas are quieter than other ducks and so they may be a good choice if the noise from your pond may disturb neighbours.

◀ *In the right light, both male and female Cayugas can appear an almost iridescent green colour.*

Muscovy

The domesticated descendants of the wild Muscovy duck are found throughout most of the developing world, originally hailing from Mexico and Central America. They were often kept as part of the farmyard scene for their meat. These are the only domesticated ducks not descended from wild Mallard.

Wild Muscovy ducks are either black or white, while domesticated Muscovy ducks can be found in many different varieties. The heads of each variety contain the identifying red crest around the eye, which in some instances can spread across the face. In the wild, Muscovy ducks fly and sleep in trees, and their domesticated cousins have retained the strong claws needed to hold their place among the branches.

◄ *Selected clones or colour forms of the domesticated Muscovy are generally treated as sub-varieties rather than individual varieties of duck.*

ESSENTIAL CHARACTERISTICS
Size: Drake 4.5–6.3kg/10–14lbs. Duck 2.2–3.1kg/5–7lbs.
Varieties: Black and White, but there are many colour variations including Lavender and Blue.
Temperament: Can live almost as feral populations. Large males can appear to be quite aggressive.
Environment: May sleep on a fox-free duck island, but when housed requires plenty of room.
Egg yield: Seasonal, can lay large batches of big white eggs in quick succession.

Rouen

In spite of claims that the Rouen had the most distinctive flavour of any duck, even when the market was dominated by pure breeds, its main commercial use was confined to supplying large, late-season ducklings at a time when Aylesburys would have been in short supply. The Rouen was bred in France for centuries as a utility table duck, although it is not really suited to that purpose. It resembles a large mallard, albeit heavier, reaching up to 5.4kg/12lbs in weight. The modern exhibition form suffers from poor fertility. However, the slightly smaller and equally attractively marked Rouen Clair breed may be a more appealing breeding option. The larger type may be kept as exhibition birds or as a statuesque addition to the garden pond, lake or stream.

◄ *The drake Rouen (left) will lose its iridescent black head and most of its distinctive markings during the summer when it moults.*

► *The drake adopts the same muted brown hues as its duck (right) when it moults.*

ESSENTIAL CHARACTERISTICS
Size: Drake 4.5–5.4kg/10–12lbs. Duck 4–4.9kg/9–11lbs.
Varieties: One only.
Temperament: Slow and ponderous on land.
Environment: Requires plenty of room on land and water and a large house with a shallow ramp to a large door.
Egg yield: Can be a poor layer of large white, blue or green eggs.

Silver Appleyard

The Silver Appleyard is a manmade British breed of duck developed in the 1930s and 1940s by well-known breeder Reginald Appleyard. It is a heavily set bird with a broad breast, erect stance, and back that slopes from the shoulder to the tip of the tail. It was intended to be a good utility bird, laying plenty of eggs and providing good meat as well as being attractive. The drakes, weighing up to 4kg/9lb, provide fine-tasting meat, and the slightly smaller females may lay 80 or so large white eggs per season. This bird is available

▶ *Miniature Silver Appleyard ducks are expected to display the same strong fawn flecking to the back and shoulders as the large version.*

in large and miniature versions. The miniature version is a recent addition, appearing in the late 1970s and 1980s, and is now far more plentiful than the original large version. The duck has an

ESSENTIAL CHARACTERISTICS
Size: Large drake 3.6–4kg/8–9lbs.
Large duck 3.1–3.6kg/7–8lbs.
Miniature drake 1.3kg/3lbs.
Miniature duck 1kg/2lbs 8oz.
Varieties: One only.
Temperament: Active and lively.
Environment: Large and miniature versions are at their best where they have room to wander and forage.
Egg yield: Large and miniatures can lay 100–150, usually white, but occasionally blue or green, eggs.

active and lively temperament, and remains alert. The female has a silvery-white body with flecks of fawn over the flanks. The male is darker with a beetle-green head. Both have yellow beaks, orange legs and feet.

Black East India

Also classed as a bantam duck, the little Black East India was described in the first book of standards in 1875 and is known from a decade earlier. The bird, later standardized as a bantam version, is expected to weigh no more than 0.9kg/2lbs. Despite its

ESSENTIAL CHARACTERISTICS
Size: Drake 0.9kg/2lbs.
Duck smaller than male.
Varieties: Black.
Temperament: If handled and treated as a pet, the breed can be quite docile.
Environment: Will want rather more room than the Call ducks that they are often kept with.
Egg yield: A good, if seasonal, layer of small to medium size eggs.

name, the Black East India has nothing to do with East India and was, in fact, developed in the USA. Confusingly, it is also known as the Labrador duck, but no relationship to the extinct bird of that name, or to the Labrador region of Canada, is known. As with all active light-breed ducks in which breeding males have a heightened sexual drive, it is best to keep a ratio of at least two ducks to one drake. True to its name, the Black East India is black in colour, with a wonderful, almost iridescent, green sheen to its feathers. The bill should be black, and eyes, legs and webs as black as

possible. Females may have white tail feathers. When kept as a pet or in a domestic environment they can be docile. The duck and youngsters make excellent exhibition birds. When kept on a pond as a breeding population, some strains can, at some times of the year, revert to a behaviour pattern not dissimilar to that of the wild mallard, and adolescent males may take to the wing in search of new mates.

◀ *The Black East India has unrivalled iridescent green-black plumage.*

Khaki Campbell

This is another manmade British breed, developed at the turn of the 19th century by a female breeder intent upon producing a bird with good all-round utility qualities. The first Khaki Campbell was created by

▲ *A Khaki female with its modified mallard plumage.*

crossing a Runner duck with a Rouen and the offspring with a mallard. The Khaki Campbell was to revolutionize early 20th-century duck farming. When, in 1901, Mrs Adele Campbell, of Uley in Gloucestershire, purchased a Fawn and White Runner duck because she had heard it had laid 182 eggs in 196 days, and crossed it with a Rouen drake, she could have hardly been aware that she was starting a process that would eventually lead to the creation of what was to become the world's best-known domestic duck. She called the product of this cross the Khaki Campbell, after its colour, and in honour of the soldiers who were returning from the Boer War in South Africa wearing the new khaki uniforms. That colour became a utility standard. Later, exhibitors chose to breed and select strains in Light and Dark forms. Over the years, both have formed the basis of some of the most successful laying strains. As with some poultry, Campbell ducks that are exhibited differ in appearance from the utility types, some of which are smaller and, in the females, have a less even colour than the exhibition Khaki. The White Campbell should probably best be seen as a separate breed.

ESSENTIAL CHARACTERISTICS

Size: Drake 2.4kg/5½lbs.
Duck 1.8–2.2kg/4–5lbs.
Varieties: Khaki, Dark and White.
Temperament: Reasonably quiet but when in lay, utility strains are intolerant of disturbance.
Environment: Likes reasonable space and access to water, but utility strains are used to being kept very intensively.
Egg yield: Good utility birds will produce in excess of 300 eggs per year, though those kept as ornamental pets may not have such prolific egg-laying ability.

Saxony

The Saxony was created in Germany in the early part of the 20th century, and is named for the region where it originated. In the 1930s, it was first bred by Albert Franz of Chemnitz in Saxony. The species was almost wiped out in World War II, but was recreated in post-war Germany. It has only recently been recognized by the American Poultry Association. It has Pekin, Rouen and Pomeranian duck breeds in its ancestry and is a heavy breed, with males weighing up to 3.6kg/8lbs. The colouring of the drake is softer than a mallard, while the female has plenty of buff in its plumage, with blue shading. The breed is a utility bird, with the female laying approximately 200 eggs per year.

This attractive, medium to heavy duck re-established the idea that a potentially useful dual-purpose utility duck could also be selected as an attractive addition to the garden

◄ *Saxony ducks lay beautiful large, white eggs.*

ESSENTIAL CHARACTERISTICS

Size: Drake 3.6kg/8lbs.
Duck 3.1kg/7lbs.
Varieties: One only.
Temperament: Easily handled, ideally suited for the garden as a pet.
Environment: Requires a largish space and area of water.
Egg yield: More than 150 large, usually white eggs.

pond. The drake's head and neck are blue. A white ring completely encircles the neck. It has rusty-red natural mallard markings, which, along with the duck's buff head, neck and chest, provides a subtle variation on natural mallard markings.

Hook Bill

An ancient breed known in Holland in the 17th and 18th centuries, the Hook Bill has a curiously shaped curved bill which may have evolved to assist foraging in mud. The breed is likely to be descended from early Indonesian ducks crossed with the local Dutch mallard population. The bird is bred in

◄ *The Hook Bill duck has a natural bib rarely standardized in other breeds.*

ESSENTIAL CHARACTERISTICS

Size: Drake 1.8–2.2kg/4–5lbs.
Duck 1.3–1.8kg3–4lbs.
Varieties: One only.
Temperament: Seems happy with other domestic ducks. Could make an interesting pet.
Environment: Enjoys a watery environment where it can follow its inclination to forage for food.
Egg yield: Egg numbers are variable. Some ducks lay well.

several varieties today, though it is mostly kept for exhibition purposes, where the shape and form of the distinctive hook bill are seen as more important than exact coloration. In fact, many

examples of the breed, which is relatively new to the show scene, seem to follow the same bibbed pattern that is so often a feature of feral populations of pond ducks.

◄ *The Hook Bill drake has an attractive natural colour pattern, but the type and shape of the bill is considered the more important exhibition feature.*

Indian Runner

While European ducks were being selected on the basis of the quality of their meat, a far lighter duck capable of laying more eggs could be found over wide areas of the Far East. The breed soon to be named the Indian Runner was first imported in the

ESSENTIAL CHARACTERISTICS

Size: Drake1.3kg/3lbs.
Duck 1.2–1.8kg/3–4lbs.
Varieties: At least eight are known.
Temperament: Edgy and sometimes excitable.
Environment: Requires room to swim and forage and has a natural inclination to run.
Egg yield: Utility strains are recorded as laying more than 280 eggs, but exhibition strains lay far fewer eggs.

middle of the 19th century, when a sea captain brought some birds into Cumbria, England, from the East Indies. These were probably typical of many of the ducks that were bred or had evolved in Malaysia and Indonesia over the centuries. Not only were they taller, lighter and possessed a far more upright carriage than European ducks, they also laid far more eggs. While they gained a local reputation as wonderful layers, they remained practically unknown beyond Cumbria and the Scottish borders before the turn of the 19th century. Yet the breed was eventually to have an enormous impact on duck breeding and farming throughout the Western world, through the

commercial success of descendants such as the Khaki Campbell.

Few exhibition ducks compare with the elegance of a slim Indian Runner. The length of the neck is expected to make up one-third of the bird's total height of 66–81cm/ 26–32in in a drake and 61–71cm/ 24–28in in the duck. Slim shoulders and a lean and streamlined head complete the outline. In a show pen nothing compares with a column of bold upright Indian Runner ducks.

◄ *A slim bird with good proportions and elegance are more important than excessive height in the exhibition Runner.*

Call

Weighing only 0.6-0.7kg/21–24oz, call ducks are known as the true dwarves of the duck world. They are noisy and companionable, and make a perfect bird to keep in the back garden. With round heads, short deep beaks, puffy cheeks, and available in an increasing range of colour combinations, they have a cute appeal and a character that becomes obvious when they are kept as pets.

Originally it was their small size and the female's loud call that made them the breed of choice for their original role as "call" or decoy ducks. Once referred to as Dutch Decoys, they are still kept in Holland in large numbers, where they are aptly named "Kwakers". Call ducks were used to assist hunters in luring larger wild birds into traps. The loud call of the small bird was known to carry some distance. Once caught, the larger birds would also call, attracting even more birds to the trap.

The new interest in Call ducks has seen many new colours being developed and bred. The large quantity of show classes for the breed, combined with increased numbers kept, means that a broad range of colours is exhibited. Several subtle variations on the wild mallard colouring seen in continental varieties of Call ducks may in the future add interest to the exhibition scene. Not every duckling hatched will have the desired show characteristics. A percentage of ducklings turn out to be long-beaked. Nearly all make friendly garden pets, though their loud and talkative behaviour may make them unwelcome in built-up areas.

Call ducks are active birds, chirpy by nature, with their plumage at its best in late autumn and early spring. Not all birds bred for show will have the desired exhibition points.

The huge increase in Call duck numbers is a result of successful breeding of exhibition birds. The breed provides would-be owners with perfect pets which, with care and understanding, can be kept in a small area. The breed learns to interact with humans more readily than any other duck.

◀ *This all-white Call duck displays the appealing characteristics for which the breed is recognized.*

ESSENTIAL CHARACTERISTICS
Size: Drake 550–700g/19–24oz. Duck 450–600g/15–21oz.
Varieties: Plenty including Apricot, Apricot-silver, Bibbed, Blue-fawn, Black, Butterscotch, Chocolate, Dark Silver, Dusky Mallard, Khaki, Magpie, Mallard, Pied Mallard, Silver, White.
Temperament: Vocal and friendly.
Environment: Small area will suffice.
Egg yield: 25–75 eggs per year.

◀ *Call ducks are ideal birds to keep if you have a small garden pond.*

◀ *As ornamental birds, Call ducks make appealing garden companions.*

GEESE

Geese make interesting and unusual pets, and there are plenty of reasons to keep them if you have the land available. Perhaps your concern is in keeping geese for meat or eggs for the table, or you may be interested in breeding geese to sell either their eggs for hatching or the resultant goslings. Plenty of people keep their birds as ornamental garden companions, since those birds that are used to being handled from a young age can make docile and amusing pets that add colour and interest to the garden all year around. Remember, however, that geese will eat vegetation of all descriptions, and setting them near a prized flowerbed will almost certainly result in serious damage. Historically, geese have been used as guard animals, since they are guaranteed to set off a clamorous racket the minute visitors appear. By nature they are strong, and can be aggressive and intimidating birds. Geese have been recorded to live to a ripe old age, so making a commitment to keeping a goose can be a responsibility.

The appearance of all domesticated geese is the result of human intervention. Wild geese have a slim build, horizontal body shape and the ability to fly. Domestic geese, on the other hand, are much more upright, since they have been bred to lay down body fat around their girth, which pushes them into a much more upright stance. Most domestic breeds have lost the ability to fly, despite their great wingspans.

Embden

Embden geese were established in Britain by the 1820s, though they are widely believed to have originated in Northern Germany and Holland. They were selectively bred with native British white geese, and soon gained a reputation as one of the most useful farmyard breeds. Crossbreeding is thought to have improved the original stock, and 50 years after Embdens arrived in the UK, they were one of the first breeds to be bred to a written standard, and were reported as being re-exported to northern Europe.

▼ *The Embden goose is a large bird which requires a lot of space.*

ESSENTIAL CHARACTERISTICS
Size: Gander 12–15kg/28–34lbs.
Goose 10–12kg/24–28lbs.
Varieties: One only.
Temperament: Bold and active, but not difficult to manage.
Environment: As good grazers they require a sizeable area of grass.
Eggs: Some strains can be good layers of large eggs.

The ganders were in demand to improve the nondescript flocks that were an important part of the village economy. These all-white ganders improved local populations without interfering with their valuable trait of auto-sexing. Male goslings are white and females are grey.

While the slightly heavier Toulouse became the favourite market bird, the Embden has probably had greater influence on traditional goose breeding than any other breed.

Commercial crosses have since taken over some of their traditional role and the Embden may no longer be as common as it once was, but for those wanting a useful, large, solid breed with an upright and defiant carriage, hard and tight plumage and strong, bold head, the Embden still takes some beating.

◀ *The Embden or Bremen goose is a massive bird, up to 1m/3ft in height, with a sturdy and powerful build. Embden geese are best kept by those with some experience of poultry keeping, since the male can be over-protective and aggressive.*

African

Hailing from China, the inappropriately named African goose evolved from a cross between the Chinese goose and the Toulouse in the middle of the 19th century. Aspects of its appearance resemble both breeds, since it has the beak knob of the Chinese and the coloration of the Toulouse. The breed descends from the wild swan goose species. A huge African gander, which may reach up to 1m/1yd tall, can be just as impressive as the largest Toulouse. This is one of the heaviest geese breeds, reaching up to 9kg/20lbs in weight, yet despite its great size, the breed is known to have a docile temperament. The head and neck are thick-set, even in the young. The head has a pronounced knob centred on top, which should be as broad as the head, and a dewlap hanging below the beak. The body tips at a gentle angle. African geese are bred for meat and eggs. Brown, grey and white varieties exist.

◀ An African gander may look intimidating but on acquaintance can turn out to be quite as friendly as any other 12kg/28lb giant.

ESSENTIAL CHARACTERISTICS
Size: Gander 9–12kg/22–28lbs. Goose 9–11kg/20–26lbs.
Varieties: Brown, Grey, White.
Temperament: Can be far more friendly than they look.
Environment: Will need plenty of space.
Eggs: Not the best of laying geese.

Chinese

Just as European breeders generally selected their birds for size and table qualities, Asian breeders developed their geese along different lines, particularly those originating in China. The Chinese breed is lighter in weight and is naturally capable of laying more eggs than the European and American heavy breeds. The pure Chinese are so different from other breeds of geese that many authorities suggest they descend from a cross that included either the wild or domesticated form of Asiatic swan geese (*Anser cygnoides*).

White and Brown-grey varieties are available, each with a long neck and identifying knob on top of the bill. The Brown-grey colour pattern of the Grey variety is not found in any other domestic variety, but is close to that found in the wild species. These geese are vocal and their loud voice and vivacious character sets Chinese apart from other geese. Two personality types appear to develop with these geese; either they become tame and used to handling, or they become ferocious birds that may take an instant dislike to a human. Many strains kept by conservationists and exhibitors are fine-boned and smaller than the

◀ Chinese geese are instantly recognizable by the knob at the base of the bill. They are vocal and loud.

ESSENTIAL CHARACTERISTICS
Size: Gander 5–10kg/11–22lbs. Goose 4–9kg/8.8–20lbs.
Varieties: White, Brown-grey.
Temperament: Feisty, will take on small dogs.
Environment: These geese are adpatable but are sensitive to extreme cold.
Eggs: Will yield at least 50 eggs per season, with 100 being known.

Chinese of previous years. Chinese can make excellent, if noisy, guards or feathered watchdogs. They rarely fly and make flexible all-rounders. Crosses between Chinese geese and other breeds are unlikely to lay well or fatten as economically as modern industrial crossbred or hybrid geese.

Brecon Buff

The original Brecon Buffs were kept on Welsh hill farms for generations before being recognized as a pure breed in 1934. This breed is unusual in that it originated in the UK. These birds were probably typical of several variously coloured regional populations. Light in bone with tight plumage and maximum flesh, Brecon Buffs are an active breed that can be expected to lay far more eggs than many of the heavier breeds. This is possibly one of the best dual-purpose utility breeds developed for farms or smallholdings. The pink bill helps to distinguish it from the larger American breed, which has an orange bill and is among the heaviest of today's geese. Brecon Buffs are hardy birds, used to free-ranging conditions, but they are in need of protection from foxes and other predators. They are kept for their meat, although they are also known to make good mothers. In a small-scale setting, the breed will quite quickly become used to handling by a familiar human.

ESSENTIAL CHARACTERISTICS
Size: Gander 7–9kg/16–20lbs. Goose 6.3–8kg/14–18lbs.
Varieties: One only.
Temperament: Active.
Environment: Requires a fair amount of space.
Eggs: Is one of the best-laying pure breeds.

◄ *Brecon Buff geese may make useful and attractive additions to smallholdings and larger gardens.*

American Buff

A large all-American goose that the Americans can be justly proud of; this breed shares a similar plumage pattern with the smaller British Brecon, which has caused some confusion in those countries where both are found. Apart from its larger size, the light orange bill and feet should help to distinguish it from the pinker shades standardized in the Brecon Buff. This is a hardy breed that generally lays more eggs and is easier to breed from than some other large breeds. As a useful and attractive goose that some American authorities have described as possessing a pleasant disposition, the breed may find many new friends among fanciers with enough room to accommodate a large and active bird. The breed makes good parents for their young.

ESSENTIAL CHARACTERISTICS
Size: Gander 9–12kg/22–28lbs. Goose 9–11kg/20–26lbs.
Varieties: Buff.
Temperament: Has a pleasant disposition.
Environment: A large active goose that requires plenty of room and, like all buff-coloured waterfowl, will fade in sunlight. Exhibition specimens will need extra shade.
Eggs: Up to 25.

◄ *The orange bill and pronounced dewlap help to distinguish this from the smaller and neater Brecon Buff.*

Pilgrim

Pilgrim geese are a lightweight breed, and one of the few types of goose where the sex of the bird can be identified by the colour of its down upon hatching. Male goslings are typically creamy yellow and females are grey. The colour difference continues into adulthood. The male turns white, while the female develops grey plumage. Both have orange beaks, legs and feet. Thought to have been developed in the 1930s, the breed is fast-growing and will become tame with handling. The females can make reliable broodies and attentive mothers. Similar self-sexing geese populations, with some relationship to the native breeds, seem to have been part of British poultry-yards for centuries. These may be the origin of strains developed in the USA in the 1930s, which hatch as grey females and white males.

◄ Pilgrim geese have a strong flocking instinct, and make sociable birds with a companionable disposition.

ESSENTIAL CHARACTERISTICS
Size: Gander 6.3–8.1kg/14–18lbs. Goose 5.4–7.2kg/12–16lbs.
Varieties: Subvariety has been described as West of England.
Temperament: Active, likes to range.
Environment: Suitable for most domestic and farm situations.
Eggs: Good layers of fair-sized eggs.

Pomeranian

The ancient German Pomeranian breed is a regular winner at poultry shows. Its popularity has increased in line with the current boom in small-scale poultry keeping. This breed could be of interest to those with time and room to accommodate a family of geese. Females are good egg layers, regularly producing 60 or so eggs per season. White birds with a grey head, back and wings are available. The striking solid-coloured saddle and head was developed in the breed's native Germany where, in spite of importations of other breeds, it remains popular. Its appeal is twofold, stemming from its usefulness and attractive appearance, with striking markings. The blue-grey back markings have been described as heart-shaped. The Pomeranian is a large bird, weighing in at up to 10.8kg/24lbs,

◄ Solid coloration of the neck and heart-shaped back markings distinguish the German Pomeranian breed.

ESSENTIAL CHARACTERISTICS
Size: Gander 8–10.8kg/18–24lbs. Goose 7–9kg/16–20lbs.
Varieties: Buff-backed and solid White forms.
Temperament: Docile but can make a noisy watchdog.
Environment: Requires plenty of room and a large house.
Eggs: Variable numbers of large eggs.

and in appearance it has a heavy build. Individuals make good guard birds. This is one of several European breeds of goose that is now becoming increasingly popular in both Britain and the USA, whereas in continental Europe, they are found in Buff-backed and solid White forms.

Sebastopol

Long trailing or frizzled feathers identify the Sebastopol goose, the only breed with this unusual feathering. The soft, curling feathers start at the base of the neck and cover the entire body; those on the head and neck are smooth. A variation is a Sebastopol with smooth breast feathers. White and Buff birds are known. The curly feathers are a result of a mutation and mean that the bird cannot fly, so it would not survive in the wild. Sebastopols are mid-weight geese, with the gander weighing in at a minimum of 5.4kg/12lbs. Females are reasonably layers producing 25–35 eggs per year. Some strains produce good table geese.

The Sebastopol came to prominence in the mid-19th century. It was sent to England from the port of Sebastopol in the Ukraine, but is known around the Danube. In Europe it has the alternate name Danubian, which accurately reflects its point of origin. Often two intensely frizzled birds will not breed

▼ Sebastopol geese are found in both frizzled and long trailing feather forms. Skilled breeders often use both forms to obtain exhibition examples.

true. The best pairing is a smooth-breasted Sebastopol with a frizzle-feathered bird.

ESSENTIAL CHARACTERISTICS
Size: Gander 6.4–7kg/12–16lbs.
Goose 4.5–5.4kg/10–12lbs.
Varieties: Self white and Buff.
Temperament: Reasonably quiet.
Environment: Long trailing feathers will need to be kept out of muddy conditions.
Eggs: Egg numbers can be variable.

Toulouse

The Toulouse is still the best-known heavy goose breed in Europe. These grey geese had been bred in the Toulouse district of France for centuries where they were used in the production of pâté de foie gras. Like the Embden, they were used to improve the British flocks. Later, they provided the accepted female partner in the classic Embden-Toulouse cross that became the basis of many farmyard flocks. The exhibition Toulouse is less heavy than its profusely

◄ The Toulouse makes an excellent show exhibit and an impressive picture when seen in a grassy paddock.

ESSENTIAL CHARACTERISTICS
Size: Gander 11.8–13.6kg/26–30lbs.
Goose 9–10.8kg/20–24lbs.
Varieties: Grey, Buff and White.
Temperament: Slow and stately. May not be as long-lived as some lighter breeds.
Environment: Needs lots of space.
Eggs: Exhibition strains can be poor layers.

▲ The thickset head and large dewlap add to the overall impression of a massive bird.

feathered outline would suggest, and it lays fewer eggs than would be commercially acceptable. There is no more impressive sight than a group of deep-keeled Toulouse resembling a galleon in full sail; as they move slowly and majestically across a grassed paddock.

TURKEYS

Turkeys are very distinctive-looking birds, with their large size, scrawny necks, ugly facial features and bulbous bodies on relatively thin legs. They are reared for their meat in many parts of the world, both on a commercial scale and, in some regions, by small-scale farmers. In the USA the male is called a tom, in Europe a stag. The female is a hen and the chicks are known as poults. The fleshy protuberance on top on the head is known as the snood, while the one beneath the beak is a wattle. Turkeys are vocal birds and, when kept as pets, have been known to be responsive and companionable animals.

Many species are not seen at exhibition partly because of their size. Turkeys are large birds and are difficult to transport to shows. In addition, their feathers break easily, making the birds appear less attractive and damaging their chances of gaining maximum points. Breeders who keep small flocks of pure-breed turkeys in order to supply a distinctive, quality product have an important role to play in maintaining a wider gene pool. Nearly all of these more serious breeders will use incubators rather than broodies in order to hatch sufficient poults for further breeding at the optimum time of year.

Black

The Black turkey, also known as the Norfolk Black and Spanish Black, is a medium-sized bird weighing up to 10.4kg/23lbs. It was developed in Europe from birds brought from North America in the 16th century, and is now widespread in Europe. In the 17th century, the breed was returned to North America where it was bred to develop the Slate, Narragansett and Bronze breeds. As its name suggests, it is covered in black plumage with an iridescent green sheen. It has a bright red wattle and yellowish

> **ESSENTIAL CHARACTERISTICS**
> **Size**: Stag 10.4kg/23lbs.
> Hen 5.9–6.8kg/13–15lbs.
> **Varieties**: One only.
> **Environment**: This is a large bird that needs plenty of space.

skin. Young turkeys, known as poults, may have white or bronze feathers as they grow, but these are lost with successive moults. This bird is relatively slow to mature, taking up to six months, and as such it is not a favoured commercial breed. Its yellow skin is also a factor in its unpopularity, since the public's preference is for smooth white flesh. Having said that, the texture and flavour of the Black are unrivalled in turkey meat.

This turkey breed is known for having a calm nature and likes a large area for free-ranging. However, it is in danger of becoming extinct because of its less commercial appeal. Unlike other turkey breeds the Norfolk Black can breed naturally. This breed has been crossed with other heavier breeds to produce a larger, more commercial turkey, and as such, the pure breed may be difficult to find.

▶ *The Norfolk Black is an ancient breed of turkey with plumage that is black offset by a red head.*

Bronze

The Bronze is a giant among turkeys. For a long time it was the main commercial breed, dominating the market for several decades from the mid-20th century. The breed developed from turkeys taken to America by immigrants from Europe. For many people, these are the typical farmyard turkey, with their iridescent bronze sheen. Later in its history, the breed divided into two strains: the Standard Bronze and the Broad-breasted Bronze. Thanks to industrial breeding for exaggerated type and the public's preference for white birds, the Bronze strains had practically ceased to exist by 1976, and only survived because of the efforts of fanciers and a few small-scale producers.

While large bronze turkeys are now bred in fair numbers for the table, most of those seen at shows are kept closer to a weight of 5.4–10.8kg/12–24lbs.

▶ *The copperish-bronze feathers terminate in a narrow black band.*

ESSENTIAL CHARACTERISTICS
Size: Stag 13.6–18kg/30–40lbs.
Hen 8–11.7kg/18–26lbs.
Varieties: Standard, Broad-breasted.
Environment: Space required.

Cröllwitzer

This is a relatively small turkey with unusual and attractive colouring. The Pied and American Royal Palm turkey breeds have similar colouring, though the Pied breed is an old breed that dates back to the 1700s, while the Royal Palm was first recorded in the 1920s.

This bird is kept for its eggs rather than as a meat bird to be fattened for celebrations, and is less likely to be kept as a commercial proposition because of this. The breed makes a good bird for a small farm or for dedicated breeders keen to show at exhibitions. It is an active breed and its lighter frame make it visually different to other commerical turkey breeds. However, its meat can rival the taste of other commercial breeds. The head is red; the neck feathers are white with a black tip, while the saddle feathers are black with some white fringing. The tail feathers, when opened out, appear to have a broad black band running across them. It can take dedicated breeders years to perfect the colour markings of the feathers on this particular breed. The breed has heritage status.

◀ *This breed has an active disposition and requires a reasonable amount of space in which to live.*

ESSENTIAL CHARACTERISTICS
Size: Stag 7.3–12.7kg/16–28lbs.
Hen 3.6–8.7kg/8–18lbs.
Environment: Space required.

Glossary

Addled A fertile egg in which the embryo died during the early stages of incubation.

Back-cross Mating an offspring with one of its parents.

Barred A plumage pattern of black and white stripes across the feathers.

Beard Feathers under the throat.

Boule An unusually thick neck with a rounded, mane-like appearance.

Blue A feather colour that usually describes blue-grey; in reality it is a dilute black.

Brassy A yellow tinge to the feathers of white birds that can be caused by weather. It is genetic when found on the shoulders of roosters of other-coloured fowl.

Broiler A fast-growing table fowl, usually mass-produced and that may be killed as young as 42 days old.

Clean-legged Having legs that are free of feathers.

Cloddy Inclined to be heavy in the hand.

Closed flock A flock of birds of the same breed kept by one breeder. Continuation of the flock is maintained by breeding birds within the group without the addition of outside bloodlines.

Cobby Having a compact, rounded body shape.

Cock A male bird that has completed one or more breeding seasons. Also known as a rooster.

Cockerel A male bird that is in its first breeding season, or one wearing the current year's leg ring.

Columbian A colour and the name of a variety of fowl. The neck hackle has a darker stripe, usually black

▼ *Light Sussex bantam hen.*

surrounded by white at the edge, or buff in those varieties described as Buff Columbian.

Columbian restriction A breeding selection which results in the black being restricted to the hackle, wings and tail. It resulted in the now iconic "Light" in Light Sussex.

Crest A bunch of feathers on the head, or in case of the Poland breed, a full globular crest of feathers.

Cuckoo Feather marking similar to, but less distinctive than, barring, as found in the Marans.

Cushion A mass of soft feathers on the rump of the female. This feature is usually well developed in the Cochin.

Dominant Of characteristics that appear in the first hybrid generation when each parent shows different characteristics, eg rose comb is dominant to single. For example, the gene for the rose comb is dominant over the one for single comb, so first-generation offspring of a rose comb-single comb cross would have rose combs.

Duck foot In a normal fowl foot the fourth or back toe is well spread out allowing stability, particularly when taking a backward step. Duck foot is used to describe a limp fourth toe (as found in ducks). This is considered a serious fault in all breeds other than Asil, where thanks to its legendary bravery it is said that the males never take a backward step.

Duckwing A plumage colour where the normal red-brown, is replaced by white, or in the case of the gold duckwings, yellow or straw colour.

Feathered legs Legs completely covered in feathers, including the outer toes as in the Cochin, or just to the outer shanks as in the North Holland Blue or Modern Langshan.

Flight feathers The large primary feathers of the wings.

Fluff The soft feathers covering the body.

Frizzled Plumage in which the

feathers are turned back.

Furnished Fully feathered; a rooster with fully-grown sickle and saddle hackle feathers.

Genes Hereditary factors carried by the chromosomes.

Genotype The genetic composition of an animal. The genetic constitution for a character which may or may not be expressed.

Hackles The long-pointed feathers of the neck and saddle in male birds.

Hen A female bird older than 18 months old.

Henny or Hen-feathered The absence of neck and saddle hackle and sickles in males, standardized in the Sebright and a variety of Old English Game.

Heterozygous A bird formed by the union of a male and female germ cells which are unlike for a given character.

Homozygous When a bird is derived from the union of male and female germ cells, each containing a factor for a specific character, it is homozygous for that character and will breed true for that character.

Hock The knee joint.

Hybrid A bird produced by a program of inbreeding or recurrent reciprocal selection; or an incross or incross breed achieved only with inbred lines of which the coefficient of inbreeding is not less than 50 per cent.

Inbreeding Mating very closely related individuals such as father to daughter, or mother to son.

Incross breeding Mating of highly inbred strains of different breeds.

Incrossing Mating of highly inbred strains of the same breeds.

Keel The breastbone.

Lacing When the edging around a feather is a different colour from the ground colour. Lacing can be single as in Silver-laced Wyandotte, or double, as in the Indian Game and Barnevelder.

Lavender A paler but true breeding form of the colour "blue".

Leader A single spike at the back of the comb. It can be long and thin leading to a point, and follow the line of the comb base as in the Hamburg, or it can follow the line of the neck as in the Old English Pheasant Fowl.

Leaf-comb A comb resembling a leaf as in the Houdan.

Line-breeding Breeding within a family but avoiding the repeated use of closely related individuals.

Mealy Describes a defect in buff-coloured birds in which the feathers are speckled with white.

Mottled Having a white tip to the end of the feathers as in the Ancona, or black spots on the legs.

Moult The shedding of the feathers prior to the growth of new plumage.

Muffled The muff and beard of the Faverolles, Houdan and Belgian Bantam.

Outcrossing The practice of introducing unrelated genetic material to the breeding pen in an attempt to add desirable characteristics to the breed. In the first crossing it reduces the chances of disease and abnormalities that can result from excessive inbreeding, but can later promote the introduction of recessive genes to a wider gene pool.

Partridge The natural wild pattern of red jungle fowl, in which the male has a black breast and red or orange hackle and the female has a light brown body. Alternatively the more complex variation found in the Cochin or Wyandotte, for example, in which the female's feathers are pencilled with concentric rings of a darker hue.

Pea comb Three small, single ridges joined at their base.

Pencilling Small stripes on the feather that may either follow its outline or run across it. Different to barring.

Phenotype The appearance of an individual; the production record of an individual fowl is termed its phenotype for the particular character.

Primaries or Primary feathers The ten feathers of the wing lying between the finger-feathers at the wing-tip plus the small axial feather that lies between the primaries and secondaries.

Pullet A female in her first year of egg laying. In exhibition terms this also means either a bird bred after November 1 of the previous year, or wearing a closed ring issued for the current year.

Pyle or Pile A partridge pattern in which white feathers replace the standard black in the male and brown in the female.

Reachy The carriage of a bird that is held aloft.

Recessive Refers to characteristics that are suppressed in the first generation.

Rose comb A broad comb in which the surface is more or less covered by small pimples or protrusions.

Saddle The back of the male bird in front of the tail.

Sappy or Sappiness A yellow tinge in birds with white plumage which arises from excessive pigmentation. It should not be confused with brassiness, which occurs on the surface of the feathers.

Scales The overlaying rows of horny skin tissues covering the legs.

Secondaries or Secondary feathers The set of quill-feathers in the wing between the axial feather and the body.

Selection Choosing specific birds from which to breed because they have a desirable characteristic.

Self colour All one colour.

Shaft The stem of the feather.

Shank The part of the leg between the hock joint and the foot.

Sickles The long curved feathers in a rooster's comb.

Single comb A flat comb which has serrations along the edge. It is held vertically in male birds and some-times falls to one side in female birds.

Spangling A spot of colour at the end of a feather that is a different colour to the rest of the feather.

Spikes The serrations on a comb.

Sport When a new characteristic appears in one generation and is transmitted to succeeding

▲ *A Game fowl.*

generations. The White Wyandotte originated as a sport of the Silver-laced Wyandotte, for example.

Spur The horn-like growth on the shank of a male bird, occasionally seen in females.

Squirrel tail The tail-feathers sloping forward over the back.

Stag A male bird, more commonly used to describe turkeys, but sometimes used by exhibitors.

Strain A group of birds constituting a family within a breed or population.

Top crossing Mating an inbred, selectively bred male with an unrelated, less selectively bred female.

Type Body conformation or shape. Breed-type refers to the shape and size of the breed. A "good type" means a bird that closely conforms to the breed standard.

Under-colour The colour of the lower part of the feather that is hidden by overlapping feathers.

Utility Poultry bred for egg production and/or for the table.

Variety Different feather colours standardized within the same breed. Occasionally it refers to two different standardized comb types.

Vulture hocks Stiff feathers that grow down from the hock joint.

Wattles The red fleshy appendages hanging from the throat at the base of the beak.

Web (of the feather) The barbs of the feather on each side of the shaft.

Wing bars A line of differently coloured feathers that appear across the middle of the wing in the plumage of some varieties.

Work or working The small bumps or protrusions that are found on the top of a rose comb.

INDEX

▼ *A Sicilian Buttercup.*

ACKNOWLEDGEMENTS

The publishers would like to thank all the poultry owners who generously brought their birds to be photographed, and to the show organisers who allowed the photography to take place. Every effort has been made to trace poultry owners of all poultry photographed and we apologise for any ommisions made to this list of contributors.

G Abraham, Will Allen, Brian Anderton, K Arnold Dennis Ash, R A Axman, Sue Baker, Keith Barnes, Frances Bassom, H A and S Beardsmore, Richard Bett, Tim Biela, S Black, Laura Bovingdon, G D Brearley and Son, K Britten, Alan Brooker, H W Broomfield, Alan Brown, Sue Bruton, T and J Buck, Callum Burney, Andy Capel, Samantha Carr, J Christopher, Len Clark, Christine Compton, K & J Cox, Robin Creighton, Emma Crook, M Crowther, Sandy Cummin, A and J Cumming, C Curtis, Angela DuPont, A J Davies, P E Davy, R Eden, Celia Edmonds, Derrick Elvey, B Evans, Mrs G Evans, R A Everatt, Carle Faiers, James Firth, Mr R Fontanini, R Francis, Mr and Mrs Frizzell, Steve Fuller, Caryl, Steve Gilliver, S Goodwin, Zoe Gracey, Lee Grant, Terry Gregory, Griffin and Gifford,

Colin Gullon, Mike Hadfield, L Hampstead, Fred Hams, Louise Hidden, Mandy and Barry Hobbs, Graham Hodge, Derek Howells, Derek Hoyland, C Hughes, David Iley, Anita James, C Joiner, J Kay and Sons, Stuart Kay, Alan Kemp, Mr Kemp, M Kennedy, Andrew Kerr, M R Knowles, P Knowles, R J Lomas, P Lutkin, Jim Marland, C Marlies, Trevor Martin, Gary McKinstry, Michelle van Meurs, Lorna Mew, Priscilla Middleton, Jason Millward, Pedro Moreira, Nigel Morgan, J Nibs, J P Oakley, J Owen, Harry Pannell, Jack Partridge, Mark Perkins, J Pickles, William Pimlott, Alan Pollington, Mr Power, David Pownall, Alan Procter, John Pummell, Craig Ramus, Robin Ramus, Antoinette Reese, John Rich, C Roberts, Richard Rowley, Brian Sands, W and J A Sharp, R and C Shepherd, Ian Simpson, Ian and L Simpson, Ian Sissons, Tony Smith, John Soper, Sylvia Soper, Tamsin Spicer, Anthony Stanway, Eddie Starkey, Clive Stephens, Margaret Stephens, Martin Stephenson, D Stone, Penny Strutt, O'Sullivan, A and J Tacey, P B Tasher, C F Taylor, N R Taylor, J Tickle, G Tinson, Mrs Troth, David Vicente, Team Wakeham, Hugh Wallace,

B G Ward, P Watkinson, N Watson, R Watts, Lesley Webdale, John and P Wilde, A Wilson, Mrs Wincott, Rodney Wood, Steve Woodcock, Sue Woods, Chris Woolley, Simon York.

The publishers and author would like to thank the following for allowing photography:

The National Federation of Poultry Clubs show at Stafford. The help of the Poultry Club of Great Britain members and judges at the Poultry Club of Great Britain. Affiliated shows run by Arun Valley Poultry Fanciers Society, Hants and Berks Poultry Fanciers Society, Kent Poultry Fanciers Society, Norfolk Poultry Club, Reading and District Bantam Society, Surrey Poultry Society.

Thanks to Forsham Cottage Arks, Mr and Mrs Raymond and Angela May, Derrick Elvey, Andrew and Lorna Mew, Jane Booreman, Priscilla Middleton, Mandy O'nions and Mark Hobbs, Neil Weller, Philip Lee Woolf at Legbars of Broadway.

Thanks to the following photographic libraries and individuals for permission to use their images: (c = centre, l = left, r = right, t = top, b = bottom) Alamy: page 54, 77, 84b. Ardea: page 57tl, 84t, 114t. Clare and Terry Beebe: page 72t, 73tc, 73tr, 109t, 109br. Corbis: page 99t. Fotolia: page 37t. Istock: page 107. Dave Scrivener: page 49b, 53br, 58t, 58c, 108b, 109bl, 117l, 117r, 120t, 121t. Tim Martin, Eternal Photography page 91b. Philip Lee Woelf: page 71.

This edition is published by Southwater an imprint of Anness Publishing Ltd
info@anness.com
www.annesspublishing.com

© Anness Publishing Ltd 2024

Publisher: Joanna Lorenz
Editorial Director: Helen Sudell
Editor: Simona Hill
Photographer: Robert Dowling
Designer: Nigel Partridge
Production Controller: Ben Worley

Fred's expanded edition, *A Complete Practical Guide to Keeping Chickens* (isbn 9780754835653), also gives advice on caring for poultry.

If you like the images in this book and would like to investigate using them for publishing, promotions or advertising, please visit our website www.practicalpictures.com for more information.

A CIP catalogue record for this book is available from the British Library.

PUBLISHER'S NOTE
Although the advice and information in this book are believed to be accurate and true at the time of going to press, neither the authors nor the publisher can accept any legal responsibility or liability for any errors or omissions that may have been made nor for any inaccuracies nor for any loss, harm or injury that comes about from following instructions or advice in this book.